The Patrick Moore Practical Astronomy Series

More information about this series at http://www.springer.com/series/3192

Myths, Symbols and Legends of Solar System Bodies

Rachel Alexander

 Springer

Rachel Alexander
Exeter, UK

ISSN 1431-9756   ISSN 2197-6562 (electronic)
ISBN 978-1-4614-7066-3   ISBN 978-1-4614-7067-0 (eBook)
DOI 10.1007/978-1-4614-7067-0
Springer New York Heidelberg Dordrecht London

Library of Congress Control Number: 2014950407

Cover illustration by Sarah Young used with permission from Walker Books UK.

Printed on acid-free paper

Springer is part of Springer Science+Business Media (www.springer.com)

Dedication

*This book is dedicated to Asra Jilani,
my best friend, and Milo Powell,
Alejandro and Luca Iglesias-Whittaker,
and Heather Morgan, astronomers
and artists of the future whose enthusiasm
is contagious.*

Preface

Astromythology is the mythology, associated language and symbolism of the main bodies of our Solar System. The term is derived from the Greek word *astron,* meaning "star." This is not to be confused with cosmology, which is the science or theory of the birth and evolution of the universe.

The term 'mythology' can mean different things to different people at different times. Many hark back to a time when celestial bodies themselves were revered, perhaps with deities then being associated with them, by the ancient Greeks and Egyptians.

A mythology is kept alive by language and memories, as myths are often stories whose original authors remain unknown, and these stories were usually relayed in the oral tradition. The myths were obviously changed, often by their many different storytellers, and were often based on even earlier mythologies. The Romans absorbed ancient Greek myth, and it is from them that we owe our planetary names as well as many of the names of their satellites. A mythology can encompass folklore, fables, stories, allegory and philosophy and can be changed to suit religious and political ideas. Language itself is an ever-changing animal that must impact on the relaying of the myths themselves.

The science in this book is intended to contextualize and co-exist with the mythology. The perceived logic of science and the perceived illogical nature of mythology, with its associated symbolism and superstition, are not natural bedfellows, but they are closer than might be expected. The planet Vulcan, for example, was imagined by scientists to try and explain the inexplicable. Science itself is constantly changing and evolving. The tracking of minute particles, neutrinos traveling faster than the speed of light, in October 2011 by scientists at CERN, potentially challenges our existing knowledge of science. (CERN in Europe is at the cutting edge of research in the field of particle physics.) In the end, what they saw turned out to be an error in measurement, but it does show that science is also ever-changing. Another example is the German-born physicist Albert Einstein (1879–1955), who at first rejected the idea that the behavior of electrons could be explained in probabilities. Science and stories can co-exist, be mutually beneficial and even enhance each other, as illustrated by names in the U.S. space program, such as Apollo after the ancient Greco-Roman god of light and the Sun.

Astromythology can be seen as an attempt or attempts to provide a narrative to the spiritual and awesome experience of looking up at a clear, night sky.

Astromythology and its associated language and symbolism really is everywhere, if you begin to look for it. It is such an integral part of the human psyche that we do not even stop to think about it. Astromythology symbolism features on flags—symbols of entire nations—is in the periodic table, is part of language, is integral in literature, including religious texts, and is found in the names of food, festivals, flora and fauna. Even modern cultural references such as the Disney cartoon character Pluto and popular Western chocolate bars are named after astronomical and astromythical terms—*Milky Way, Mars* and *Galaxy*—in addition to popular tabloid newspapers such as *The Sun* and the *Daily Star*.

The word *planet* originates from the Greek and means "wandering heavenly bodies." Planets are differentiated from fixed stars, as they appear to have a route of their own. Planets do not produce any light of their own; they merely reflect the light of their nearest star. The classical planets (Venus, Mars, Mercury, Jupiter and Saturn) were known to the ancients. From as early as the sixteenth century B.C., court astronomers were appointed in ancient China to watch and follow the activities of the heavenly bodies. The ancient Chinese allocated to each planet an element. Mercury was given water; Venus, metal; Mars, fire; Jupiter, wood; and Saturn, earth. By 800 B.C. the astronomers of Babylon and China had some early knowledge of the movements of the planets. The ancient Greeks and Romans believed that the planets themselves were living deities, revealed in the names chosen to represent them. The astronomer Johannes Kepler believed that planets had souls and explained their orbital movements by conjecturing that they were being pushed around by angels (Fig. 1). Onions were cultivated by ancient Egyptians.

Fig. 1 Angels pushing around the planets

They worshipped them on altars, as it was their belief that they looked like planets, and they swore oaths by onions.

The ancient Greeks believed that the Sun and Moon were also planets. They held the belief that Eurynome, goddess of everything, had assigned the planets, with a Titan and Titaness deity, to rule over them. The order went as follows: the Sun was ruled over by Theia and Hyperion; the Moon by Atlas and Phoebe; Mars by Dione and Crius; Mercury by Metis and Coeus; Jupiter by Themis and Eurymedon; and finally Saturn by Rhea and Cronus.

Astromythology shows its relevance to everyone in the names of the days of the week, with Monday being the Moon's day. Tuesday is associated with Mars, more obvious in the French language, where Tuesday is known as *Mardi*. Wednesday, *Mercredi* in French, is associated with Mercury, and Thursday, *Jeudi* in French, is associated with Jupiter. Friday, *Vendredi* in French, is associated with Venus, while Saturday and Sunday are obviously the days of Saturn and the Sun.

This book sets out to incorporate mythology and symbology concerning the main players in our Solar System, our Sun, the planets and their moons, and to begin to explore the etymological, political, and philosophical connections with astromythology.

Exeter, UK   Rachel Alexander

Acknowledgements

With thanks to:

NASA, Exeter Cathedral, Mars UK, John Watson, Maury Solomon and all at Springer Publishers.

Marek Kukula, Public Astronomer at RMM, Greenwich

Elizabeth Cunningham and Greg Smye-Rumsby at the Planetarium Greenwich.

Asra Jilani, without whom this book would not have existed.

Vivienne Morgan, Annette Iglesias, Oliver Whittaker, Shez Ziauddin, Atiya Hassan, Hamid Syed, Suzie Morgan, Michael Cole, Jonathan Dymond, Davina Powell, Jill Ashby, Aoi Yabuki, Tomoko Ishikawa, Mum, Mary Buckell, Mike Richardson, Paul Attwell and everyone that encouraged me, believed in me and helped me.

About the Author

Rachel Alexander grew up in Nottingham, England, the centre of the Robin Hood legends. She studied English and History at Hatfield Polytechnic, UK, including ancient Greek history and philosophy, before completing her post-graduate Certificate of Education (the UK qualification for Teaching) at Exeter University. She then went on to teach English in a secondary school, before becoming a Learning Support Tutor at Exeter College. She has taught English to Japanese students every year in Exeter and/or Dorset, and has taught English to Japanese students in Nara, Japan. Following her time in Japan she returned to education and studied Fine Art at Plymouth University. In this course she specialized in Portraiture and researching and compiling information on symbolism, which led to and has enhanced her interest in mythology. She has had a long-time interest in amateur astronomy.

Contents

Chapter 1

Solar Heroes and Sun Gods

Our bright, beautiful, powerful light and heat-giving Sun has provided the inspiration for countless heroes, leaders and gods. The mythical English King Arthur is one of these, along with his father, Uther Pendragon, and the sorcerer, Merlin.

King Arthur of Camelot is indisputably a British national hero. He is the most popular and romantic of the mythical Sun heroes. As the son of Uther Pendragon he is closely linked to Saint Michael, Lord of the Light and a warrior angel named in the Bible. Arthur is also known as the Lord of the Summer, associated with fertility, light and the Sun in opposition to Morgan Lefay, or fate, a winter goddess of darkness and death.

Arthur's famous round table is round like the Sun, and his magnificent sword, Excalibur, which is sheathed in a magical scabbard and which protected Arthur from being wounded, is supposed to shine with the light of many torches. It is as if the sword personifies Arthur himself, shining the light of truth, justice and goodness. Handmade swords all have an alchemic quality about them, as earth, air, fire and water were all used in their creation, especially the element of fire and its association with the Sun. The English King Richard the Lionheart, in the twelfth century, also wielded a sword that he claimed was Excalibur, immediately associating himself with King Arthur. (Arthur was believed to have been a real king in medieval times.) King Edward I of England also had a round table made with 24 seats around it which is now in Winchester Great Hall. King Arthur was regarded as the model of kingship.

After the end of the Wars of the Roses, Henry Tudor, who was Henry VII, moved his court to Winchester, as this was thought at that time to have been the sight of the mythical Camelot. On September 20, 1486, the first Tudor heir was born at Camelot and so was appropriately named Arthur. But Prince Arthur died

© Springer Science+Business Media New York 2015

R. Alexander, *Myths, Symbols and Legends of Solar System Bodies*, The Patrick Moore Practical Astronomy Series, DOI 10.1007/978-1-4614-7067-0_1

prematurely and suddenly at age 15 in 1502. Even today in England the next heir to the throne, Prince Charles, has 'Arthur' among his middle names.

Excalibur is supposedly the magical sword that was given to Arthur by the Lady of the Lake. The sword is also known as Caliburn, Caladfwlch and Caladbolg, the latter name linking it to the Irish hero Fergus and other Irish heroes. The name Caladbolg comes from words meaning "hard" and "lightning."

The idea of Arthur's Knights of the Round Table could have originated from the Fianna or Fenians of Ireland, a large group of soldiers formed around 300 B.C. with the purpose of protecting the High King of Ireland. It is also believed that many Arthurian legends are similar to those of the famous Irish hero Finn Mac Cool and his Fianna.

There are many, many stories of King Arthur. The English poet Edmund Spenser (1552–99) wrote an epic poem, *The Faerie Queen,* whereby Arthur was involved in adventures in Fairyland before he became king of Britain. Geoffrey of Monmouth's Arthur battled with the Romans, while the later Arthur of Sir Thomas Malory's (1405–1471) *Le Morte d' Arthur* was the product of Uther Pendragon disguised as the Lady Igraine's husband by the magic of Merlin. Young Arthur was taken away from Igraine and was brought up by others. He became king when Merlin magically pushed a sword into a stone and spread the word that whoever could withdraw this sword from the stone would become king. The Round Table appears in this epic prose as a wedding gift from Guinevere's father, its round shape important as there was no obvious head of the table.

Arthur left Britain to battle against Lancelot (who had been Queen Guinevere's lover), who had taken her abroad with him. Arthur left his son, Mordred, to rule Britain in his absence. Mordred rose up against his father, forcing Arthur to return to stop his wayward son. Arthur and Mordred were supposed to have had their last battle on Salisbury Plain. Arthur killed his son but was mortally wounded. This final battle is known as the Battle of Camluan to the Welsh. Arthur has also been claimed by the Irish. He is the king of Britain's son and dared to steal the dogs of Finn Mac Cool and Bran. The Fianna chased Arthur to Britain, took back their dogs and got Arthur to pledge his loyalty to Finn Mac Cool. Finn is one of the most famous Irish mythological heroes. The Druid Finegas educated Finn.

Finegas caught the legendary Salmon of Knowledge, which Finn cooked. During the cooking, Finn famously burned his thumb. He sucked his thumb and acquired the salmons' knowledge. Finn, like Arthur, is not dead but sleeping, waiting to someday wake up for Ireland in its time of greatest need.

The age of King Arthur's reign is known as the Golden Age. Arthur is supposed to have been buried at Glastonbury Abbey. However, according to one legend, he healed from his wounds and lies in a deep sleep waiting for the bell to ring, which is supposed to wake him up when he is needed most, which makes him the once and future king and sound somewhat Christ-like. Some people believe that the island of Bardsey is Avalon, the final resting place for King Arthur. According to some Arthurian myths, Arthur's ship is at the bottom of Bardsey Sound. Arthur himself is either living as a hermit or sleeping in a cave on the island. Bardsey is supposed to be the place where 20,000 saints are buried and was established in the sixth century by Saint Cadfan as a place of pilgrimage. Legend has it that if you are buried on Bardsey, you will achieve salvation, because it is a gateway to another realm. In Welsh tradition, the dead set out for this island.

The mythical King Arthur has always been associated with the Sun. He emerged from the darkness and the obscurity of a relatively humble upbringing, like the Sun emerging from behind clouds or from the imagined darkness before the creation of light. Arthur is supposed to have been at his best in battle during the waxing of the Sun, his strength and fighting ability receding along with the Sun's rays. Obviously, Arthur's power shone, increasing in the light of the knowledge of his noble birth as son of King Uther Pendragon, Head of Dragons. His obscure upbringing adds to the air of mystery, as did his sword, Excalibur, believed to have been forged from metals found in a meteorite, metal that had fallen from the sky, a gift from the gods.

King Arthur, if he existed at all, is believed to have reigned in the sixth century, a long time after the Romans had left Britain, conveniently in the historical period known as the Dark Ages, where there is very little written evidence, as most people were illiterate. Obviously, this makes it very difficult to prove or disprove his existence.

Centuries later, his actual existence is arguably insignificant. It is what he represented that matters. He became the symbol of stability, harmony, balance, continuity, bravery and a good and just leader. He is what all leaders and benign monarchs aspire to be.

King Nuada in Celtic mythology is similar to King Arthur in some ways. Like Arthur, he was a just and fair king, wise and gallant but a force to be reckoned with on the battlefield. He was king of the gods and, like Arthur, was the proud owner of a magical sword. His was named Freagarthach, which was always terrifyingly accurate, always killing enemies. He, too, was a solar hero, as king of the Tuatha De Danaan, a dynasty of deities referred to as the gods of the light in Ireland before Christianity.

King Nuada fell from grace when he unfortunately had his arm sliced off in battle against the enemy, the Firbolg. The Tuatha De Danaan were victorious, but Nuada could no longer retain his position as king or leader, as his body was now incomplete. He was succeeded by Bres the Beautiful, who was an unstable and imperious ruler. He did not inspire others, as Nuada had done, and was perceived as weak. With a new enemy, the more fearsome Formorii, the wicked gods of darkness who were about to attack the Tuatha De Danaan, what was needed was the stability and sense of order that the former King Nuada had represented, again like Arthur. Bres, Nuada's replacement, was half Formorii himself, so he was not exactly trustworthy. In fact, he behaved predictably by joining with the Formorii against the Tuatha De Danaan. A magical doctor, Dian Cecht, created a replacement arm for Nuada from gleaming silver so he was whole again, and therefore could be king. Eventually, the magical doctor's son used more magic to replace the metal arm with a real one, so that Nuada could lead the Tuatha De Danaan into a war against the dark Formorii gods.

In the Christian religion, Jesus Christ, the son of God, is the supreme solar hero. He is described as the 'light of the world,' and his location as a baby was illuminated by a star that directed the three wise men to him. He was also illuminated by the Holy Spirit when John the Baptist baptized him. Like our Sun provides us with light, Jesus is believed to do the same, only his light is the grace and holiness of God.

Helios was the classical Greek Sun god, who crossed the starry firmament of heaven in a chariot driven by four white horses to provide light and traveled down in the evening into the sea, where he sailed around Earth at night in a bowl that shone like a god. Helios was the son of the Titan god Hyperion. Helios had three children, including his son Phaeton and daughter Circe, who are also connected to astromythology.

One day, Phaeton decided that he wanted to drive Helios's chariot, as he thought that it looked exciting. Unfortunately, this proved too much for the young and inexperienced boy. He lost control, causing Libya's sands to turn barren, and parts of Africa to become arid. Zeus intervened, without a moment to lose, steadying the runaway horses and narrowly preventing Earth from becoming an incandescent, flaming ball. Phaeton was unable to keep his balance, though. The impulsive boy fell from the chariot into the sea, where he drowned.

Circe was a demi-goddess and an enchantress. Some might say that she was a witch. She was also the lover and captor of Odysseus, the hero of Homer's *Odyssey*. She took great delight in changing men into animals. She turned Odysseus's men into pigs and Picus, son of Saturn, into a woodpecker, since he refused to be seduced by her. Out of jealousy, she cast spells that turned poor Scylla's legs into hellish, canine creatures before turning her into a headland of rock (Fig. 1.1).

Fig. 1.1 The companions of Odysseus after enchantment by Circe. (Illustration by the author)

Fig. 1.2 Picus as a woodpecker, punishment for refusing Circe. (Illustration by the author)

Picus was a very attractive young king of Ausonia. He was not exactly short of potential wives or lovers but was extremely fussy and refused them all. Eventually he fell in love with a beautiful nymph named Canens, who sang so sweetly that her music melted his hardened heart. One day, Picus was out chasing boar when Circe spotted the gorgeous young king. It was lust at first sight for Circe, but Picus had eyes for Canens alone and turned down Circe. Used to getting her own way, she felt angry and humiliated by Picus. In her rage, she turned the unlucky Picus into a woodpecker (Fig. 1.2). Canens was heartbroken.

The statue known as the Colossus at Rhodes, one of the Seven Wonders of the World, was supposedly of the deity Helios, but it was destroyed by an earthquake around 226 B.C.

Helios's sister—Eos or Dawn—was supposed to climb aboard her chariot, drawn by horses called Lampus and Phaethon, when night ended and ride to Olympus to announce the approach of her brother Helios. When Helios appeared, she became Hermera and accompanied him on his travels until, as Hespera, she announced their arrival on the western side of the Ocean.

Helios became involved in the story of Orion the Hunter when his sister, Eos, became enamored of him. Orion was irresistibly good-looking. One day he met Merope, daughter of Oenopion, and liked her so much that he asked permission from her father to marry her. However, the father's blessing came with a condition. He had to rid the island of Chios of the many wild creatures that were potentially harmful. Oenopion demanded proof, so Orion had to bring him skins and furs from the creatures each night. Orion, being a very skilful hunter, completed his task and came to claim his bride. However, Oenopion refused to keep his promise. Orion was understandably disappointed and so furious that he got drunk and claimed Merope anyway by having sex with her before falling asleep. Then, while Orion was defenseless and vulnerable, the cruel Oenopion gouged out his eyes and dumped him near the sea. Following a prophecy from the oracle, he somehow managed to travel in an easterly direction. Eventually, his path led him to Eos. It was love at first sight for her. She persuaded her brother to give him back his sight.

Sol is the Roman name for Helios. Sol Invictus, meaning "unconquered Sun," was introduced by the Emperor Aurelian in 274, who credited his victory over the Palmyrene Empire to this god in 273. Aurelian tried to set up Sol Invictus as the main god in the Roman pantheon. Sol was very popular among the soldiers of the Roman army and remained so until their conversion to Christianity under the Emperor Constantine in the fourth century. Some Roman coins had the Emperor Constantine's face on one side and the god Sol on the other, linking the Sun with the light brought to the world by Christianity while keeping a foot in the camp of the pagan pantheon.

The Greco-Roman god Apollon/Apollo is the god of light, prophesy, healing, the arts and the intellect. His origin is uncertain, with one theory connecting Apollon to the Indo-European *apelo*, which means "strength." Another theory links Apollo to Appaliunas, a god from Anatolia, whose name means "father lion" or "father light."

Apollo, one of the twelve Olympian gods and twin brother of Artemis, was later known as the Sun god, as revealed in the tragic story of Clytie, the Sunflower Maiden. Clytie was a beautiful water nymph with long golden hair that glistened in the Sun. She used to watch out for Apollo's chariot to come through the gates of heaven every morning when the world was flooded with the sunrise. She fell in love with the strong, handsome god. At night she returned to the water and dreamed of the beautiful, shining Apollo. She became consumed by her love for him, unable to think of anyone or anything else. She stopped returning to her pool each night and waited for his return in the morning. She longed for him to notice her and return her love, but she waited in vain. For 9 days, golden Clytie watched Apollo, eating and drinking nothing but her own bitter-tasting tears. By the ninth day, she was unable to move. Her feet were rooted to the ground, her arms and fingers were green leaves and her lovely face with its halo of golden hair had become a flower—a sunflower. Clytie could no longer speak nor cry but still turned her head towards the Sun, as do all sunflowers to this very day. In a variation of this myth, Clytie was transformed into a heliotrope and fell in love with Helios, not Apollo (Fig. 1.3). In ancient Greece, a heliotrope was any flower that followed the Sun.

In another myth, Apollo saw Cupid playing with his bow and arrow and laughed at him. However, Cupid declared that Apollo wouldn't be able to harm him with his arrows (he was immortal), but he, Cupid, could easily hurt Apollo with his. Cupid shot an arrow at Apollo and then fired a second one, with a lead tip, at the lovely young woman, Daphne, daughter of a river god. When the first arrow struck Apollo,

Fig. 1.3 Clytie as a heliotrope, or sunflower. (Illustration by the author)

he fell in love with Daphne. However, having been shot by the lead-tipped arrow, for Daphne, the feeling was not exactly mutual. Apollo asked Daphne to marry him. She refused. Daphne was strong, beautiful and active and loved to run and hunt. She ran away after yet another proposal. Apollo, fierce and fueled by love, pursued her.

At first, she thought that she could outrun him, but when she realized that he had almost caught up with her, she asked her river-god father to change her form, so that Apollo would not know her. Daphne soon felt her feet rooted to the ground and could no longer move them. Her outstretched arms became branches, and around her body grew the bark of a tree; her hair and fingers became leaves waving in the wind (Fig. 1.4). She had been changed into a laurel tree! He kissed the tree and declared that he would not have harmed her and that he would wear a crown of laurel leaves, as would all men when they were crowned conquerors. He also announced that her leaves would always be green, and they are to this day. In a variation of this myth, Daphne was spirited away to Crete, where she became known as Pasiphae.

Apollo also became infatuated by the beautiful sybil of Cumaea. (Sybils, or Cassandras, were the female equivalent to male prophets.). He wanted to have sex with her. She agreed on the condition that he would grant her a wish. She wanted to live for a 1,000 years. This wish was granted and became a perfect example of the warning "Be careful what you wish for, as you might just get it." She did live for her requested millennium, but began to shrivel up as the years passed by. She was imprisoned by her aging body rather like a genie in a bottle until all that remained was her voice.

Fig. 1.4 Daphne transforming into a laurel tree to escape Apollo. (Illustration by the author)

Apollo had many lovers, including Coronis, the Muse Thalia, Pythia, Aria, Cyrene and the nymph Dryope, with whom he had children.

Apollo was a protagonist in Homer's *Iliad*. Agamemnon neglected Apollo's priest, so Apollo caused a deadly plague on his army in retaliation.

The U. S. space program that was called Apollo was named for the Greek Sun god. The *Apollo 13* mission badge depicted three golden horses pulling the Sun across the sky.

King Louis XIV of France lead such an extravagant lifestyle that he was known as the "Sun King" when he took to using Apollo's golden Sun emblem to symbolize his perceived power and status. Some of Apollo's power could have been seen to have rubbed off on him as he reigned from 1643 to 1715, the longest-reigning king in Europe. In the Mongol Empire, Genghis Khan was believed by some to have been the child of the Sun himself, making him a solar hero.

Claiming descent from deities adds credibility or kudos, instills fear and reinforces power. Some Icelandic people have claimed descent from powerful Germanic gods, such as Thor, god of thunder and lightning, and Odin, chief of the gods. Alexander the Great and his family claimed descent from Heracles/Hercules, himself a son of Zeus. Arrian wrote of Alexander's birth in his *The Life of Alexander the Great* that he believed that something more than human was involved in his birth.

Despite Arrian writing considerably after Alexander's death in the second century and his work being based on questionable evidence, the myth of Alexander as being semi-divine still lives on. In Greek mythology, Theseus claimed descent from Poseidon and Ion from Apollo. The ancient Greeks believed that character was destiny, as Heraclitis (535–475) had stated, believing that men should develop in their own way even if they were not free to shape their own destiny. In the twenty-first century we regard Theseus, who slew the minotaur, and Alexander the Great (356–323 B.C.), who defeated the Persian Empire in 331 B.C., among other incredible feats, as heroes in their own right, so divine ancestry would merely add to their reputation.

The Egyptian pharaoh was believed to have been an incarnation of the Sun god, Ra, when he was alive. Sometimes, the Sun god disguised himself as the reigning pharaoh to have sex with his queen, so that he could produce his son, the next pharaoh. This allowed him to reinvigorate the solar genes, his genes being more powerful than the diluted ones of the pharaoh. When the child was born, it was revealed to Ra in the temple for him to receive and validate his divine child.

Ancient Egyptian mythology is among the most complex. Different places worshipped different gods, and gods were worshipped during different historical periods. If that were not complex enough, some gods changed their names.

The following story involving Ra, the Sun god, is of special interest, as it was found in the pharaoh Tutankhamen's tomb in the 1920s.

Ra was disappointed in people, who had forgotten that he made the world. Some people were even scheming to remove him from power. He asked all the other deities what he should do about this. They overwhelmingly advised him to punish his ungrateful subjects. Ra's eye metamorphosed into a bloodthirsty, fierce lioness called Sekhmet. Sekhmet was commanded to kill all the people on Earth, a task that she took great delight in executing. By the end of each day, she had brutally slaughtered a large percentage of Earth's population.

Ra decided that he no longer desired to kill all the people, but Sekhmet had the taste for human blood and did not want to stop. Ra realized that he had to trick Sekhmet to prevent her from annihilating the remaining people. Ra ordered the men to collect a specific, red-colored stone and the women to brew thousands of containers of beer throughout the night while Sekhmet slept. They all worked very hard without questioning Ra's authority. Then the men ground the red stones into a fine powder, which was then combined with the beer, making it look like blood. The huge quantity of liquid was then poured onto the earth, where it became a blood-red lake. Sekhmet drank so much of the blood-colored alcohol that she became very drunk. She returned to Ra's palace without killing anyone. The next day, Sekhmet awoke with a ferocious hangover, but had forgotten all about her murderous mission.

The people were all too aware that Ra had saved them. Their love for him was renewed, and they did not plot to overthrow him any longer.

The god Re or Ra, the Sun god, was often combined with Amun, the hidden god. Re embarked on a large, complex and arduous journey across the sky in a boat, and, from the West, he traveled for a further 12 h in the night-boat. This was not as simple or straightforward as it might sound, although Re had other gods to help him through his adventures, such as when he came up against the terrifying snake, Apep, and the forces of confusion.

In one version of the mythology, the Sun god, when he has sunk, becomes Osiris, personified as a man with the head of a ram. His pilot-gods navigate his boat for a single hour. The dead come alive again, apparently temporarily resurrected by the air from Earth that Osiris has bought with him.

When the Sun reaches the next stage of the Duat, or realm of the dead called Urnes, it is confusingly known as Af Ra. Here, the boat of Af Ra meets up with the boats of Osiris and his attendants. The dead are given food, light and fresh air at this stage, too. The Sun fights with the snakes called Hau and Neha-her while in darkness. When they are defeated he is taken to the field of the grain-gods to rest for a short time. There, he listens to the prayers on behalf of the dead made by those still living and accepts their gifts and offerings. He proceeds to travel across all 12 parts of the Duat. He is then led to Ahmet, another part of the Underworld.

In the sixth part, the dead Egyptian kings and the spirit-souls known as the 'Khu' reside. Here, Af Ra turns to the East (as opposed to previously traveling from south to north) to the Mountain of the Sunrise. The troublesome and evil snake, Apep, blocks his pathway in the seventh section, at which point other gods and goddess, including Isis, join him. His accompanying gods stab Apep's body. His boat is towed through the eighth section by gods, but it sails unaided through the ninth. He passes over lakes in the next two sections, with his way lit by a circle of light surrounded by a snake in the eleventh section.

In the twelfth section is the enormous celestial water, or Nu, where Nut, who represents the morning, resides. The boat has to travel through the giant snake Ankh-neteru's body with Af Ra emerging out of its mouth, where it has been changed into Khepera, who is towed into the sky and led before Shu, child of Nu and god of the atmosphere. Shu puts Khepera in a gap in the half-circular wall marking the end of the twelfth section. From here, he appears to humans as the circle of light with which we are familiar.

In Egyptian mythology, the god Atum was the original local god of Heliopolis in Lower Egypt. Atum apparently brought himself into existence, appearing from the waters of Nun, the most ancient of the gods, in the form of a heap of earth. From this, he rose upwards, becoming Re-Atum, the Sun. The Sun is also supposed to have been devoured, along with the stars, by Nut, goddess of the sky from whom they emerged each morning. In another myth, the Nile Goose laid the egg of the cosmos where the Sun, Aman-Ra, was hatched.

Heliopolis, meaning city of the Sun or city of Helios, was the major place of Sun worship. Benu, the solar god of rebirth (represented as a heron), Atum, the creator and solar god, and Re/Ra were also all worshipped there. The Scarab beetle is a symbol of the Sun and its route across the sky for the ancient Egyptians.

The pyramid represented the ascent of the Sun in the sky for the ancient Egyptians, although many other civilizations had also built such structures. Over 12 different civilizations have built pyramids. Those in the Mayan and Aztec civilizations were built for ritualistic purposes. For the Tucume peoples in the Lambayesque Valley in Peru, these pyramids were solid structures. It was like they were trying to replicate mountains, which they believed to have magical defensive powers. However, for the Hattushan Hittites, who had ruled what was mostly Turkey and Syria, the pyramid had to be huge, symbolic of the status, influence and authority of their empire.

Pyramids originated from a hill symbolizing the primordial hill of creation from where the Sun god appeared. The triangle that points up energy and fire to the heavens or sky is associated with the Sun.

For the Egyptians, pyramids were grand and elaborate status symbol burial places. Their pyramids were once crowned with a smaller pyramid, frequently gilded to shine in the Sun, imitating the daily rebirth of the Sun. It is also believed that the great pyramids of Giza in Lower Egypt were lined up with the stars on Orion's Belt.

The summit of the pyramid represents the pinnacle of achievement. Step pyramids, such as that of King Zoser at Saqqara near Giza, represent the believed cosmic framework and the rising up of the Sun into the sky. For Plato, the pyramid was used to represent the element of earth (with other three-dimensional shapes representing the elements of fire, air, water and ether).

The magnificent pyramids of Giza were originally even more awe-inspiring, as they were encased with a fine, white limestone. However, this was mostly removed and used to rebuild the city of Cairo after its destruction by an earthquake.

For the Egyptians written communication, or hieroglyphs, held great power, and from the end of the Old Kingdom (2650–2450 B.C.) to the Middle Kingdom (2040–1750 B.C.), the walls of the buried chambers of royal family members were emblazoned with these hieroglyphs, known as 'pyramid texts,' which contained magical charms and spells intended to protect the dead person and to assist in the journey to the afterlife. The coffins themselves also contained magical, protective text. Builders of these amazing tombs incorporated devices such as false doors and secret passages to deter grave robbers. Many graves were entered and valuables stolen in antiquity, so these devices were largely unsuccessful. For the ancient Egyptians their afterlife was a place called the Field of Reeds, where dead souls could enjoy spending eternity with Osiris, the god of the Underworld. This Field of Reeds was an idyllic place with a wealth of crops and a happy life.

When the mummified body was taken to its tomb, priests conducted the mystical 'Opening of the Mouth' ritual. They believed that the Ba, the spirit or soul of the dead, left the body through the open mouth. Then another ritual was performed involving goods that the dead person would require in the afterlife. For the Egyptians, the dead were judged. Their heart was weighed against the Feather of Truth. It is believed that the jackal-headed Anubis placed the heart upon the scales, and the ibis-headed Thoth, god of wisdom, recorded the judgment.

Alexander the Great conquered the Persians, who ruled Egypt, in 332 B.C. He was thought to have been a demi-god, rather like Hercules. He is known to have visited the oracle of Jupiter-Ammon in Libya to prove to himself that he was the offspring of Zeus/Jupiter rather than Philip of Macedon. The name of the oracle is interesting in itself, being a fusion of Roman and Egyptian deities. Alexander founded the city, Alexandria, that bore his name. This became famous as a city of learning. A city of learning was symbolic of Alexander's power and wealth. Learning and knowledge were highly prized, and wisdom was equated with power and advancement of civilization. Alexandria had become a center for the collection of wisdom and learning. Knowledge could be both intellectually and spiritually illuminating, a commodity with comparable value to crops or even gold.

This city of illumination and learning was a grand vision and included a library where texts, including the famous *Iliad, Odyssey* and even the Hebrew Bible, were copied by hand onto papyrus. Unsurprisingly, Alexandria became a place where thinkers could meet and inevitably advancements were made, and it provided

Alexander with yet another legacy, perhaps influencing some later attitudes towards learning. This center of learning united the Greek love of learning with the Egyptian love of written communication; after all, Alexander was tutored by the great thinker Aristotle.

The importance of the Sun is perfectly illustrated in the myth of Icarus and his father Daedalus. Daedalus made wings for himself and his son to escape from Crete. Daedalus warned Icarus not to fly too close to the Sun, as it would melt the wax holding the feathers onto the false wings, and to follow him. They both left the island of Crete, soaring off then, flapping their wings like birds in the sky. Eventually Icarus began to soar upwards, close to the Sun, enjoying the power and freedom of flight. When Daedalus looked back, he could no longer see Icarus— only feathers floating upon the waves in the sea. The heat of the Sun had melted the wax on Icarus's wings, and he drowned. Icarus was guilty of hubris by underestimating the great power of the Sun and paid the ultimate price, unlike his much wiser father, Daedalus, who lived to tell their tale.

Jason was a beautiful man and was therefore compared to Apollo, making him a solar hero, too. Jason's quest was to find the Golden Fleece, literally a fleece that shone like gold. This fleece was a gift from the gods and had to be retrieved from the end of Earth. The fleece came from a ram, the product of Neptune's (in the shape of a ram) rape of Theopane, and was guarded by a dragon who never slept (Fig. 1.5). The fleece had great financial worth but had magical powers, too, including the power or ability to heal and to end famine, which rendered it priceless, worth the extremely long and difficult journey, both physical and spiritual, to obtain it.

The Golden Fleece also represented the return to power of the rightful branch of the family. Pelias was the current king of Iolcus. Pelias was half-brother to Aeson, Jason's father. He said that Jason could have the throne of Iolcus only if he could obtain the Golden Fleece, a seemingly impossible challenge, involving a journey to Colchis, which was believed to be in modern Georgia, a complex and potentially treacherous journey in itself at that time.

Fig. 1.5 The protector of the Golden Fleece. (Illustration by the author)

Jason had assistance in his task from Medea. Medea was the daughter of the king of Colchis, who owned the Golden Fleece. She was also a sorceress and granddaughter of Helios, the Sun god. Her assistance came via Hera and Aphrodite, who made her fall in love with the brave and handsome leader, Jason. He loved Medea for 10 years, and together they produced three sons. However, Jason then left Medea and fell in love with another princess, Glaucus. They did not live happily ever after. Medea was furious and caused the unfortunate Glaucus to be burned alive in a wedding costume woven with poison. Medea also murdered her own sons by Jason. She escaped in Helios's enchanted chariot to the city of Athens. Jason was left all alone and was, understandably, deeply miserable. He died when a large piece of decaying wood from his previously magnificent ship, the *Argo,* fell off and hit him, killing him.

Dragons themselves are solar creatures, as fire comes out of their nostrils. By Christian times, however, the mythical creature was seen in the West as a representation of Satan, often being associated with the dark (as opposed to the light of God). These dragons tend to have serpent-like tails, which, when tied into a knot, indicate that the creature has been vanquished, as seen in many artists' representations of the Saint George and dragon story. Saint George is often linked to Michael, the Archangel protector in the Bible, and is often depicted as slaying a dragon. Saint George, patron saint of both Georgia, South Caucasus and England, slew the dragon. Some variations depict George as killing the dragon in Egypt, whereas in others, George was a Roman soldier who was killed for his beliefs, turning him into a Christian martyr. He is often seen wearing the armor while slaying the dragon and saving the king's daughter in the process. The cross of Saint George—a red cross upon a white background—is symbolic of the soldier's victory. In Tblisi, capital city of Georgia, a somewhat naïve painting of Saint George and the dragon appears on a wall in Saint Nicholas's church. Icons of Saint George and the dragon are everywhere in Georgia, and in England, April 23 is Saint George's Day, although there are no national celebrations to mark this (Fig. 1.6).

Fig. 1.6 The legend of Saint George's slaying of a dragon guaranteed his fame. (Illustration by the author)

Fig. 1.7 Dragon from the Welsh flag. (Illustration by the author)

In Celtic religions, dragons are traditional emblems. One was pictured on the flag of Macsen Wledig and also on the Welsh national flag. In medieval times, dragons and many other seemingly fantastical creatures were believed to have been real.

Marduk was the Sun god of Assyria and Babylon. His symbol is the dragon, as he fought a long and difficult battle with the terrifying salt-water serpent Tiamet (Fig. 1.7). When he sliced her slain body in half, it became Heaven and Earth. Tiamet's eyes became the beginnings of the famous rivers, the Tigris and the Euphrates.

In Egyptian mythology, the dragon is a representative of Osiris, god of the dead and of fertility. The ancients believed that this was the dragon that was responsible for the flood of the river Nile each year. Without the river Nile, Egypt would simply not exist. It was the floodwaters that made its soil so rich and fertile. If flood levels were as little as around 6 ft or lower than average, the people of Egypt would starve.

Apep was a serpent in Egyptian mythology that symbolized chaos and disorder. Apep took great delight in trying to defeat Ra, the sun God, on a daily basis when he traversed the sky in his boat. Reassuringly, he did not manage to defeat the mighty Ra, no matter how cruel and spiteful he was.

Serpents and dragons are often interchangeable symbols. The snake is an obvious phallic symbol, and therefore can be associated with the Sun and the Moon. Snakes can bite or strangle their victims fatally but also have the amazing ability to renew their own skins and therefore represent rebirth and resurrection. In the East, dragons tend to be more positive and respected creatures than in the West.

In Greek mythology, there was a famous but terror-inspiring serpent, Python, that apparently begot itself, like the ouroboros. It enjoyed scaring people until Apollo stopped it. The impressive hero and god of the Sun and light killed the slimy tyrant with 1,000 arrows. This victory was celebrated by Apollo with his establishing the Pythian Games, an event similar to the Olympic Games consisting of an every 4-year series of competitive music and sporting events.

Gold is linked to the Sun. The Mexican Aztecs thought of gold as the excrement of the Sun god Huitzilopochtli, or remains of the Sun itself, traces being threads in

Fig 1.8 A lion represented the heavens in some ancient cultures. (Illustration by the author)

the earth. It has been associated with rulers and kings throughout history. The Spanish conquistadors in 1541, led by Francisco de Orellana, believed that a mythical city of gold actually existed somewhere deep within the rainforests of the Amazon. It was supposedly an affluent city with a king who covered his body with golden dust before washing this golden coating off in a hallowed lake. The king then delivered offerings of gold into the lake to the gods. The king was called El Dorado, the gilded one. The washed-off gold became the sacrifice. For a short time, the king or chief would have looked like a gleaming, supernatural being before being transformed back into a mere mortal. This name was also given to the legendary city, which was never found.

Gold was available to the Aztecs in the sixteenth century, but only the upper layers of society were allowed to wear gold jewelry. The nobility blatantly wore bigger, bolder, more elaborate gold jewelry, trying to outshine each other in a rather vulgar display of wealth. Gold symbolized the crop maize, the Sun and the power of the gods and therefore was used as a sacrifice to the gods. The Aztec Empire under the leadership of Montezuma or Montezuma II was annihilated by the Spanish conquistador, Herman Cortez (1485–1547), who plundered Aztec treasure, sending it back to Spain to impress King Charles II.

The Aztecs believed that everything was pre-ordained. The Spanish invaders were thought to have been the descendants of the Sun itself, according to Aztec prophecy. There are many myths involving the Sun being either swallowed or caught. These include a Celtic myth where a wolf swallows the Sun, the Sky Father, at night. In Germanic myth, the wolf Skoll caught the Sun between its strong jaws

Fig. 1.9 The Raven was believed by the Inuits and Chukchi to have put the Sun in the sky (Illustration by the author)

and proceeded to swallow her whole, but not before she had given birth to a daughter, the new Sun. This new Sun would rise up from the sea after Skoll had swallowed her mother. She would provide much needed light and warmth to the new Earth. In another Germanic myth, the cruel and fierce wolf Fenrir broke free from its magical ribbon that had bound him at the final battle, swallowed the Sun and was then killed by Odin, father of the Norse gods.

Apart from the obvious association between the Sun and the lion's ray-like mane, golden fur and incredible strength, in ancient times, the heavens were personified by a lion that swallowed the Sun every evening (Fig. 1.8).

There is a myth from the Congo that involves stealing the Sun. The hero was Mokele, whose own unusual birth involved theft. He was apparently born from an egg, like a bird. This egg had been taken or stolen from one of Wai's wives by an old woman, who could not have children. When he entered the world by breaking open the eggshell, there was darkness everywhere. Mokele decided to steal the Sun to bring some light. He did manage this and brought it to his community. He married Bolumbu and together they produced a son called Lonkundo. Lonkundo eclipsed his own father's theft of the Sun and became the hero of the Congo's Mongo-Nkundo people. Lonkundo had a dream where he, too, stole the Sun. He did not actually steal the Sun but did marry the Sun goddess, Ilankaka.

In Arctic Inuit mythology, it was believed that the Raven was the creator of the world. In ancient times, the world was in darkness. People had neither seen nor experienced light until the Raven reported seeing light in the East. The people begged the Raven to fly off to search for this. He flew off and eventually arrived at a village, where he soon discovered that the village chief had some kind of golden ball of light within his home.

The Raven managed to persuade the chief's son to play with the golden ball. He then could easily take it from him. The Raven then took the ball of light back to his own people. This was not such an easy task as it looked or seemed, and the Raven dropped the golden ball of light, which then became the Sun. The Chukchi people of Siberia have a similar myth. Here, the Raven also stole the Sun, but this time from heaven by managing to conceal it in his mouth. The Raven's son, Tangen, tickled him so much that he could not stop laughing and spat out the Sun, where it rose upwards into the sky (Fig. 1.9).

For the Polynesians their trickster hero, Maui, wanted to capture the Sun in order to slow it down. Maui enlisted the help of his brothers, and together they constructed a rope made from flax to tie into a loop or noose ready to throw upon or lasoo the Sun. After months of traveling to the East, they waited until the Sun began to rise at dawn. Then they traveled by the dark of the night so as not to forewarn the Sun. When the Sun began to rise at dawn, Maui and his brothers managed to throw the noose over the startled and frightened Sun, who unsurprisingly, struggled and tried to break free. The harder she tried to escape, the tighter the noose became. Maui then hit the poor Sun over and over again with his magic jawbone (a gift from one of the gods). He literally beat the unfortunate Sun into submission. Maui released her after a while, but his plan had worked. Instead of taking a mere seven and a half hours to travel around Earth, it now took the battered and bruised Sun 24 h. In a variation of this myth, Maui secured the Sun with a rope made from his sister's long hair. He refused to release the Sun until she agreed to shine for longer in the summer and for a shorter period of time in the winter. The island of Maui is named after the hero. It is believed that Maui caught the Sun over the volcano Mount Haleakala, which means "house of the Sun."

In some mythologies, the Sun is portrayed as either a person in human guise or with human traits or qualities. The West African people of Benin and Togo had their creator gods, who produced twins, one of which was the male, Lisa. He kept watch over the East, the Sun, the daytime and the sky and was believed to have shown people how to utilize metal.

For the Celts, the greatest warrior was the Sun himself, the vanquisher of the darkness. The Celtic Sun god was Lugh. The golden flowers of the gorse are symbolic of the energy of the Sun and of the Sun god, Lugh. It is believed that because it flowers throughout the year and remains evergreen, gorse contains a droplet of Lugh's life-giving energy. Lugh was a radiant warrior and sometimes appeared in battles to help heroes. In the Cattle Raid of Cooley, he helped the hero of Ulster, Cuchulainn, who just happened to be his son (his mother being the mortal Diedre).

The queen of Connaught, Medb, was the proud owner of a beautiful white bull. This impressive animal accidently escaped and ended up among the king of Ulster's cattle. Medb wanted another impressive bull and had her eye upon the brown bull of Cooley. Medb's men offered to buy her this brown bull, but threatened that they would forcibly remove it if their offer was not accepted. This infuriated the men of Ulster, who declared war. Cuchulainn, the hero of Ulster, singlehandedly fought a 100 men from Connaught, bravery that could be expected from the son of a god as important as Lugh. In one version of this myth, Cuchulainn was wounded at one point in the battle. Lugh himself appeared. He healed his son and made him sleep while he held off the opposition. In another myth, Lugh is supposed to have

Fig 1.10 Inti, the Incan Sun god. (Illustration by the author)

knocked out the only eye of the king of Fomoi with a stone from a sling. Lugh also fought on the same side of his son, Cuchulainn, in the Tain War. Lugh, being a Sun god in human form, had an enchanted spear that he was able to break suddenly into flames. Lugh was last seen almost 2,000 years ago, when the high king, Conn of the 100 Battles, claimed to have glimpsed him through an enchanted mist, where he made some predictions about Conn's future.

The Inca Sun was depicted in human form as a golden circle or disc with a face on it. He was also known as Inti. Inca royalty saw themselves therefore as 'children of the Sun,' and emperors were regarded as living personifications of Inti and the Sun. they built solar observatories, the oldest of which, at Chankillo, Peru, is believed to be 2,300 years old (Fig. 1.10).

The Slavic Sun god is depicted as a beautiful young man, sometimes as born anew and dying each day. In Zoroastrianism, Mithra was known as the god of light and the Sun. He was depicted as surrounded by rays of light. He later morphed into a savior hero character resembling a young man wearing battle costume, including a tunic and helmet decorated with signs of the zodiac.

The idea of the cult of Mithras was brought back to Rome by soldiers, eventually being connected with the Roman Sun god Sol Invictus, or the 'unconquerable Sun.' In Chinese mythology, the goddess Xi had ten sons who were all Sun gods. The sons were only permitted to shine in the sky on their designated day of the week. (The ancient Chinese week had 10 days.) However, the mischievous young boys decided to disobey the rules and all shine on the same day. This had devastating consequences for Earth, causing a massive drought.

It seemed as if Earth was dying, so something had to be done. The Emperor Yao prayed to the god Di Jun, who decided to get his marksman, Yi, to sort out this problem. Yi surprisingly shot nine of the sons from the sky, leaving the one surviving boy as the Sun. When each son was shot down from the sky and died, his soul or spirit fell in the form of an enormous crow. Earth then recovered from the terrible drought. The Sun-sons were believed to be the cause of the behavior of impulsive children, with no thought for the consequences of their actions.

In Hopi mythology, the Sun dresses up, and so seems somewhat human. Just before it rose up in the East, the Sun put on a gray fox skin, and then it became dawn. This was known as the white dawn. When he was the bright morning yellow dawn, he removed his gray fox outfit, substituting it for a yellow fox outfit.

There is an Aztec myth about a rivalry for the position of the Sun, which assigns the candidates and brothers, the White Tezcatlipoca (known as Quetzalcoatl) and the Black Tezcatlipoca (known simply as Tezcatlipoca), with rather unattractive human qualities. Tezcatlipoca became the first Sun, but Quetzalcoatl decided that he wanted to be the Sun that was made from stone. Tezcatlipoca, understandably, was furious and caused his jaguars to eat all the people. Quetzalcoatl became the new Sun and created new people. Tezcatlipoca retaliated by changing these new people into monkeys. Quetzalcoatl was devastated by this and resigned as the Sun and created a hurricane that moved the monkey people out of the way. Then T-laloc became the third Sun, but he eventually abandoned the position. Then Huitzilopochi assumed the role of the Sun. He shone incredibly brightly. Huitzilopochtli sometimes sent dead people's souls back to Earth as hummingbirds.

There are some mythological examples of when the laws of nature or the usual solar routine is disrupted or changed in some way. Zeus stopped both the Sun and Moon from providing light so that Gaia could not find a magical herb that would protect her monstrous children, the giants. This enabled him, along with the help of Heracles/Hercules, to win the War of the Giants. Zeus also reversed the laws of nature to ensure that King Thyestes of Mycenae resigned in favor of Atreus. Helios, the Sun god, turned his chariot around towards the dawn, causing the Pleiades and other stars to retrace their courses and the sun to set in the East. In another myth, the Sun was so love-struck over the beauty of Leucothoe that he rose too early, set too late and sometimes didn't bother to shine at all. Occasionally, he made the days longer during the winter.

In the Bible, God made the Sun stop in the middle of the sky, delaying it from sinking for a full day when he listened to Joshua, revealing that God was on Israel's side. The Moon was also stopped. The day was lengthened by God's intervention, the additional daylight allowing Joshua and the Israelites to destroy the five Amorite kings and their armies.

In Celtic myth, the Dagda, chief of the Celtic gods, held the Sun in place for 9 months while Boann, goddess of the Boyne River, was pregnant by the Dagda. This divine intervention was used to prevent Boann's husband, Nechtan, the owner of the Well of Knowledge, from discovering the infidelity of his wife.

In the Old Testament of the Bible, the Sun is considered to be one of two great lights in the sky. There is no actual association to a specific sky god with the Sun, although some Roman mosaics portray Jesus surrounded by either sunbeams or a solar halo, in a solar chariot. This links Jesus to sky gods such as Helios and Apollo.

The Celtic god Belanus, meaning 'bright one,' was connected by the Romans to their light and Sun god, Apollo. The Celtics fire festival of Beltain on May 1 was named after Belanus. It is also known as May Day. It was commemorated by the lighting of bonfires, or the Fires of Bel. During the Beltain festival, Celtic people worshipped Bel, Belensu or Bile, the god of both life and death. He could be regarded as a solar god, as he was victorious over the dark and brought the people nearer to the next harvest. The second Fires of Bel, which occurred when they were relit by torches, symbolized the Sun's rays and a new start. This is linked to the Moon, as Beltain is a Moon festival, as are other myths.

The Slavic Sun god was called Dazhbog. He is supposedly married to Myesyats, the Moon. Their arguments were believed to have caused earthquakes, as they were so tempestuous.

In a Cherokee myth, the Sun, in her younger years, had a lover who would only come to her at night. There was, however, one condition to this arrangement—that the Sun could not see his face. She agreed to this, but was naturally curious as to his true identity. One night, she marked his face with dirty ash. The next night, she was truly horrified and ashamed when she discovered that her own brother, the Moon, had a dirty ash-covered face. Her brother was filled with remorse and shame at being exposed and stayed as far apart from the Sun as was possible.

The Sun was jealous of her celestial rival, the Moon. She noticed that people, when they looked up at the Moon, smiled with admiration and awe. However, when they looked up at her, they could only squint, as her rays shone so intensely. The Sun felt full of envy and resentment and burned with even more ferocious heat. The people could not cope with this extreme heat and started to die. To fight back, survivors made the merciless decision to murder the Sun, intending to ensnare her when she called at the home of her daughter at midday.

The first few attempts were not successful, resulting in an unfortunate man being transformed into a snake with toxic blood. The snake also had the ability to kill just by glaring at his intended victim. Rather inauspiciously, this snake murdered the Sun's innocent daughter. When the Sun found this out, she was inconsolable and hid herself away. This resulted in the world having no light or heat. The people attempted to reclaim her daughter from the Darkening Land, but failed. The Sun was devastated when she heard about this, and her continuous sobbing caused floods here on Earth. She only smiled again when she heard a drummer playing. It was at this point that the world gained its heat and light back again. The Sun's heat stayed constant after this tragic incident, but people still have to screw up their eyes to look up at her.

In one Canadian Inuit myth, the light was removed. This time it was stolen. A boy who was wearing a raven skin complete with feathers flew up to the sky, which was light. He then fertilized the daughter of the Chief of the Skies by transforming himself into a cedar leaf, lying in a stream. The girl swallowed the leaf when she drank from the stream and became pregnant. She gave birth to a boy. This boy was permitted to play with a bladder that contained daylight. Eventually, the boy put his raven skin back on and flew back down to Earth with this bladder of light near the Nass River. There, he asked the people for fish, as he was hungry. They refused to give him any fish, which made the boy angry. The raven then threatened them with breaking the bladder of light if they did not give him a fish. The people merely laughed at the raven, so he broke open the bladder and daylight seeped out through it and into the world.

Some solar myths are connected with divisions of time, especially the division between night and day. The Slavic Sun god Dazhbog was born again every morning, growing increasingly older on his trip through the sky, resulting in him reaching old age by evening. In another myth, the gods provided a giant by the name of Day a chariot and a horse to pull it. Day was charged with leading them around Earth every day and night. The giant's horse was named Shining Mane, as he illuminated Earth and sky with the light from that glorious mane.

Another Sun god was Utu (also known as Shamash) in Mesopotamian mythology, who traveled each morning through the eastern Mesopotamian mountains. The gods allowed the gates of the sky to be opened, where Utu/Shamash remained throughout the day, shining for the people on Earth. Then he journeyed back via the Underworld. Often, he would guide the dead to the Underworld and occasionally would return spirits to Earth to mingle with the living. He directed oracles or prophets and was also the god of fairness. Finally, he was omniscient.

The Egyptians explained the division between night and day by the Sun god's previously mentioned complex journey across the sky and through the Underworld.

In New Guinea, an extremely unpleasant character, the son of Earth and the Sky called Geb, kidnapped human children before decapitating them. After a while, the local people rebelled over this abhorrent behavior and, more importantly, over the loss of their offspring. They retaliated by catching Geb and decapitating him. The head escaped in an easterly direction, under Earth, eventually emerging at Kondo, where the Sun rose. There, Geb's severed head climbed somehow onto a curly yam stem up and into the sky, eventually becoming the Sun itself. It then traversed the sky in the opposing direction, eventually coming back down to Earth and beneath it back to Kondo. It has continued to travel this same route every day since.

The Aztecs offered still-beating hearts to the gods, which they believed would provide the Sun with energy to move across the sky, having lost its strength through the nocturnal journey through the Underworld. The Hindus explained the division of night and day in terms of Sunrise, Noon and Sunset by the three steps that Vishnu, a Vedic Sun god, made when he walked across the universe. In Norse mythology, both dwarves and mountain giants were terrified of daylight and therefore of sunlight, and so were creatures of the night and the dark. The mysterious but amazing rock formations, such as the Riesengebirge, were explained in mythological terms by giants that had been caught out at sunrise and were therefore turned to stone or rock.

Light is obviously associated with the Sun but also with beauty, intellect and spirituality. The famous classical Greek philosopher Socrates (469–399 B.C.) was physically regarded as ugly but was believed to shine with an inner beauty. Light is associated with creation. God made light in the Bible, and in Germanic myth sparks were used to create the Sun, Moon and stars. Light is associated with enlightenment both spiritual and intellectual, with haloes and light rays revealing deities and those who have been chosen for a specific purpose such as the apostles, saints or the Virgin Mary. The Sun depicts direct knowledge while the Moon represents indirect knowledge, like its indirect, reflected light.

Chapter 2

The Sun as a Powerhouse

The Sun is a majestic, orange, gaseous ball, a nuclear fusion reactor that glows incandescently. It is a vast 865,000 miles (1.4 million km) in diameter compared to Jupiter's diameter at 88,650 miles (142,984 km) and that of Earth at 7,909 miles (12,728 km) across. It is 93 million miles away (150 million km) but still manages to retain a phenomenal influence on all the planets, planetoids and their satellites in our minute part of the universe.

The Sun is the star whose gravity holds our Solar System together, although it wobbles a little when spinning on its axis due to the influence of the gravitational pull of its planets, especially Jupiter. The Sun's light consists of all the hues of the rainbow, giving us a bright and varied palette of color. The light of the Sun is so bright that it can blind us. We have to squint to look at it with the naked eye. (NEVER try to look directly at the Sun, though, not even squinting. It will likely cause eye damage and even blindness.) In the daytime, its brightness obliterates that of other stars, which are still up there in the sky.

It is amazing to think that as far back as the fifth century B.C. the pre-Socratic Greek philosopher Anaxagoras (500–428 B.C.) believed that the Sun was a molten mass. He also knew that it was enormous, many times bigger that the Peloponnese.

The Sun was born around five billion years ago. Our Sun, like all other stars, was born in a nebula, an enormous stellar nursery. Nuclear fusion brought our star into existence and is what keeps it and the system of planets, asteroids and moons it supports alive. It has existed for around a third of the time that the universe itself has. It is now just another middle-aged yellow star that will eventually explode in the distant future, becoming a white dwarf.

When a star reaches the white dwarf stage, it shrinks. Our Sun will become small, with an extremely dense core. Its active life will be over at this stage, with

© Springer Science+Business Media New York 2015
R. Alexander, *Myths, Symbols and Legends of Solar System Bodies*, The Patrick Moore Practical Astronomy Series, DOI 10.1007/978-1-4614-7067-0_2

only a core of carbon and oxygen remaining. The Sun will continue to shine very faintly while it cools down, eventually fading into the darkness of the universe. Its likely destiny will be as a considerably smaller, dead star, a cold black dwarf. White dwarf stars are somewhat elusive. The nearest one is Sirius, the Dog Star's companion, the very small Sirius B. This was discovered in 1862 and is referred to as 'the Pup.'

We can look forward to discovering more about the Sun in the near future as ESA's (the European Space Agency) solar orbiter will get closer to the Sun than anything else has. It will have a heat shield at the front, as it will face temperatures of up to 500 °C. It will contain instruments that will be able to capture images of the Sun. This is supposed to be ready for launching in 2017.

The Sun exerts its influence on our planet in many ways. An obvious way is that it provides the heat and light that green plant life requires for the process of photosynthesis. The areas near to the equator receive the most sunlight on our planet, and this is where we have lush, green rainforests.

The Aurora Borealis, or northern lights, are the result of the atmosphere of Earth and the Sun colliding. Aurorae can also be found on other planets, such as Saturn and Jupiter, and on Jupiter's moon Ganymede; they have very recently been discovered on the planet Uranus.

Coronal mass ejections (CME's) and solar storms can disrupt people's lives here on Earth, changing our planet's magnetic currents and causing the electricity powering entire cities to shut down temporarily. In 1989, a solar storm affected Earth's magnetic field, which, in turn caused a power failure in the city of Montreal, Canada. The Sun's activity can also affect national power grids and disrupt air travel and satellite navigation.

The Sun is certainly awe-inspiring and has been deified by many peoples and in many belief systems. Our tiny area of the vast universe is named after it—our *Solar System*. When The Sun expands in the future as it becomes a red giant, it will engulf Mercury, Venus and possibly Earth itself. The Sun provides us with the appropriate conditions for intelligent life to thrive here on our planet, but could eventually become the cause of our demise.

The idea of a heliocentric cosmos is not particularly new. The ancient Greek Aristarchus (310–230 B.C.) suggested that Earth was just one of the wandering stars (along with Mercury, Venus, Mars, Jupiter, Saturn and the Moon) that traveled around the Sun. This system did not make sense to the ancients. Aristarchus used spherical orbits. The sphere is a symbol of perfection, eternity and the vault or firmament of heaven. The circle is a symbol of the cosmos and the supreme god and the eternal, as it has no beginning or end.

A geocentric model of the then-known cosmos was devised by the Greek philosopher Aristotle in the fourth century B.C. and then by Claudius Ptolemy in the second century A.D. This was not seriously challenged until the Polish astronomer Nicolaus Copernicus in the sixteenth century proved mathematically that our Solar System is heliocentric rather that geocentric. Copernicus kept his workings quiet, and his text *On the Revolution of the Heavenly Spheres* was not published until the year of his death, in 1543. The Sun then became both a political and religious topic. Copernicus's text was placed on the Inquisition's forbidden book list, and the Italian astronomer Galileo Galilei was placed under house arrest for supporting

Copernicus's ideas. The concept of a heliocentric system rather than a geocentric one would have been regarded as heresy, as it contradicted Psalm 93 from the Bible and therefore the Church: "Thou hast fixed the earth immovable and firm."

Galileo was one of the first people to look at the Solar System through a telescope (that he made himself), so he could prove that Earth revolved around the Sun empirically, which did not please the Church and challenged its authority. The geocentric system was what people were used to. The Exeter clock, donated by Bishop Peter Courtenay around 1485 to Exeter Cathedral in Devon, England, shows a geocentric system with the Sun (the fleur-de-lys) and the Moon revolving around Earth.

Exeter Clock

The Periodic Table of Elements, devised by the Russian Dimitri Mendeleev (1834–1907) includes helium, believed to be the second most plentiful element in the universe with the atomic number 2 and the symbol He. Helium, which our own Sun produces from hydrogen, is named after Helios, an ancient Greek Sun god. Norman Lockyer discovered this element and gave it its name, as he thought that it could only be found on the Sun. He chose the suffix 'ium,' as he assumed that the new element was metallic. The chemist William Ramsay was able to reveal that helium was an inert gas, not a metal, and suggested that the name 'helion' was more appropriate. This suggestion fell upon deaf ears, though, and the name helium remains to this day.

Many divisions of time are connected to the Sun, the obvious solstices and Midsummer's Eve, the night before June 21, which is the longest day. This is supposed to be a time of especially potent magic, as the Sun's power is at its highest point before its gradual waning. In the Christian religion, this time is known as Saint John's Eve. In popular folklore, it is believed that if it rains on Midsummer's Eve, this will have a negative effect on the nut harvest. In ancient times, pagans would light fires, the idea behind these being to help the Sun's power fight off malevolent spirits for a bit longer. Some jumped through the fire, with the dual purpose of supposedly cleansing the soul and protecting the crops. They also tried to copy the Sun's journey through the sky by burning cartwheels made from straw, which were pushed down a steep hill. Today, in the twenty-first century, some people still gather at Stonehenge at this date each year.

Midsummer's Day itself is when the Sun's power wanes, an abundant time for malevolent spirits. Midsummer's Eve is believed to be a popular time for covens to meet. Superstitions connected with this date can seem somewhat bizarre. Walnut trees on this day were believed by some to have been gathering places for mischievous spirits. If a girl wanted to dream of her future husband, she could pick the yarrow flower from a young man's grave or tomb and put it beneath her pillow on this date. Also, if a girl picked a rose on Midsummer's Day and wore it when she attended Church, the lucky man who took it from her was supposed to become her future husband. There were also other superstitious customs and rituals that, if enacted on this day exactly, were believed to make a woman fertile or to see which local people would die in the next year.

Celtic people divided the year into eight parts, each with a festival to revere the Earth Mother. One of these festivities was the summer solstice, which included the longest day, June 21. For the Celts, this was a night for celebrations welcoming the light and the return of more darkness. Celts today might commemorate this day by meeting up with friends and walking in a circular route or with candles, or by throwing on a fire a piece of paper with obstacles in their life written upon it. They might also make an altar to the Sun on which they could lay flowers and make a wish.

The Celtic people also celebrate the winter solstice, also known as Yule, Midwinter or the Return of the Sun. This is the beginning of a new period of more daylight. They might create a wheel or wreath made from evergreens, which symbolize eternal life. Wreaths are still used for decoration in homes at Christmas. The winter solstice was when the Sun reached its furthest point south, the shortest day. The Celts thought that the Sun stood still for 12 days at the time of the solstice, around December 21. The yule log was burned to provide light on the darkest of winter days.

Celts also celebrate the spring equinox (March 21–22), the first day of spring, which is when day and night are of equal length. This division of the year includes Oestre, or Easter. Activities for the Celtic people mainly include being outdoors, planting seeds and spring cleaning. However they might also try to walk along leylines, also regarded poetically as dragon paths, when their energy is at its greatest. Many of our Easter traditions involving eggs originated from Celtic rituals.

Some Christians in ancient times believed that the Sun danced with happiness early on Easter Sunday morning in celebration of the rising from the dead of God's son. It was at this stage that the lamb and a flag might be seen on the surface of the Sun.

Christmas and Easter also have connections with the Sun. Our Christmas Day—December 25—was not officially given as the birthday of Jesus, the son of God, until 440. Christmas rather conveniently incorporated the Roman pagan Saturnalia festival and the winter solstice. At the winter solstice, people celebrated the victory of light over darkness, and the word Easter derives from 'Eostre,' the Anglo-Saxon goddess of the dawn.

According to folklore, midnight is known as the witching hour, when ghosts, ghouls, goblins and other magical creatures were most likely to appear. It was also the time when contact with spirits was supposed to be more successful, the optimum time for séances and ghost hunts. Midnight was supposedly the time of the highest position of the spiritual Sun, connected with spiritual knowledge.

A 13-month year was observed by peasants and country folk in parts of Europe for a long period after the Julian calendar was instituted. The superstitious notion that the number 13 is unlucky comes from the number of the Sun's death-month. This superstition remains to this day.

The Julian calendar was named after the Roman Julius Caesar Augustus, who introduced this in 45 B.C. to replace the more complex Roman calendar. Julius Caesar employed a Greek astronomer and mathematician known as Sosigenes of Alexandria to create a new calendar, and he came up with a variation of the 365-day-year Egyptian solar calendar but with an extra day every 4 years.

New Year is celebrated in many cultures, albeit at different times. It represents cosmic regeneration, renewal and the increasing power of the Sun. The Sikh festival of Vaisakhi is the holiest day of the year and is a vivid, global festival. It involves feasting, dancing, color and martial arts and a ritual that involves the Sikh flag being taken down from the flagpole. The flagpole is washed and a new flag is attached to it. This symbolizes a time of new beginnings and the commencement of the new solar year.

In folktales Nameless Day occurs when events occur in 'a year and a day.' It is associated with the sister of King Arthur, Morgan Le Faye. The explanation for the day in folklore is that the king of the old year is dead and the king of the new year has not yet emerged. Morrigan, the destroyer goddess of death, had to be placated by people fasting on this day so she would allow the Sun to return. The Sun was essential for the New Year to occur and the cycle of time to continue.

It is also believed that the Sun is associated with bird migrations. Birds are thought to use the position of the Sun and the stars to help them to find their way. Sometimes birds have been known to pause in their migration on cloudy nights.

The Mayan people of the pre-classical period looked up at the clear night sky and used astronomical observations to create their calendar, which ended on December 21, 2012. This has led to speculation about the end of the world. The Mayans were competent mathematicians and astronomers who invented the concept of zero as a number and recorded the movements of the heavenly bodies on tree bark 'paper' using a hieroglyphic writing system. We actually do not know what was supposed to occur after December 21, 2012.

The Mayans did not notice that the position in the sky of both the Sun and the Moon affected the weather here on Earth and fertility, especially with crops such as corn. They invented calendars using heavenly activity to provide divisions of time. They particularly noticed the Sun's movements. Their year began on July 16 and, like our own year, was divided up into 365.25 days, with the next year beginning on Day 366. These days were divided up further, like our weeks. There was also another calendar of 360 days, subdivided into 18 short months. These calendars co-existed. The calendar's invention was attributed by the people to their creator god Itzamma, as was their writing system. Their gods were revered and feared. The Sun god was thought to have been the father of humans and the Moon deity was believed to be their mother. Their deities were all connected to nature-like gods of the sunrise, and Chac, the deity in charge of thunder, lightning and rainfall. The gods were feared, as they were supposed to have annihilated and recreated human-kind on more than one occasion. Human sacrifice was used to honor and appease their deities. The Mayan people could successfully predict some events, such as solar eclipses, and some have attributed to them apocalyptic prophecies.

The Sun, being the largest spatial object in our Solar System, does affect time itself. The nearer something is to a large object such as the Sun, the more spacetime is bent and warped. The Sun has bent spacetime, causing a blip in the orbit of the small planet Mercury. The weight of the Sun has actually slowed down time itself! This is known as the Shapiro Time Delay, discovered by the American astrophysicist Irwin. I. Shapiro (1929-).

Flags were originally an emblem of a god or ruler, borne into battle as a sacred symbol of supremacy. They now represent an entire nation. It is interesting to note that a significant number of national flags have some kind of heavenly body symbol—Sun, Moon and/or stars emblazed on them, evidence of how astro-mythological symbols have become part of the human psyche.

Almost all of the Australian/southern Pacific flags contain astro-mythological symbols, while few European flags contain them. Only countries near the Black Sea and South American islands or countries bordered by the sea have flags containing astro-mythological symbols. The flags of Argentina and Uruguay contain a specific type of Sun, the Sun of May, symbolic of the resistance to the Spanish in 1810, important to these countries as part of their journey on the road to independence. Flags of Antigua and Barbuda and that of Malawi include the Sun above the sea and a 'V' shape upon the former flag that celebrates the victory of independence from Britain. The Mexican flag includes the eagle and the cactus, dating from 1325. The Aztec people were told to settle where the eagle landed on a cactus in their own history. The Sun appears on many flags, sometimes representing royalty. The rising red Sun in the top band of the flag of Malawi comes from the coat of arms of the earlier colony of Nyasaland.

The flag of Niger, West Africa, contains an orange, circular Sun, as does the flag of Bangladesh, which is a green field representing the green and fertile land and the red circle, which represents the Sun rising and the bloodshed to achieve independence for the country, which started to use this flag in 1972.

The Sun is politically significant on the flag of the Philippines. It has a Sun, whose eight rays are symbolic of the eight provinces that rebelled against Spain.

On the flag of Namibia, devised in 1990 after its independence from South Africa, is a small golden Sun in the top left-hand corner. This Sun also has 12 triangle-shaped rays that symbolizes the Sun itself and the power and energy that it provides. The Sun is represented on the 1997 flag of the Solomon Islands by a thin yellow diagonal stripe through the center of the flag.

The Japanese flag, dating from the nineteenth century, is a white field containing a red circle representing the Sun. The Japanese call their country 'Nihon,' the Land of the Rising Sun. The Sun is an emblem of Japan, as is the chrysanthemum, which

Fig 2.1 For the Japanese, the chrysanthemum represents both the Sun and their nation. (Illustration by the author)

is an emblem of the Japanese Imperial House because of the ray shape of its petals, symbolic of the Sun (Fig. 2.1).

The Japanese Sun goddess, and one of the most important Shinto deities, is Amaterasu, meaning "the one that possesses the great Sun." Amaterasu originated from the left eye of Izanagi, the male part of the primeval couple. She was in conflict with her brother, the ruthless storm god, Susano-Wo. They set out to prove which deity was the most omnipotent. The deity that could produce male deities would triumph in this conflict. Susano-Wo produced five male gods from the fertility beads Amaterasu wore in her hair and on her arms. (The male deities were produced from both Amaterasu and Susano-Wo.)

Both deities thought that they had triumphed in the conflict. Susano-Wo celebrated his supposed win by storming Earth, causing violence, destruction and general mayhem. Amaterasu was deeply afraid and hid herself away in a cave. Being the Sun goddess, this caused Earth to be shrouded in total darkness. The other gods wanted her and particularly her light back, so they managed to tempt her out of her hiding place with a wild dance and a mirror. Amaterasu caught a faint glimpse of her own reflection, and as she emerged from the cave to see more of her own beauty, the world was once again illuminated by the Sun.

Until 1954, Amaterasu was worshipped as a holy ancestor of the Japanese royal family, and the mirror was included within the imperial insignias. Amaterasu is still worshipped at Japan's most important shrine, the Grand Shrine of Ise. Her grandson's grandson, Jimmu Tenno, became Japan's first emperor. Japan's emperors have all claimed they were descendants of Amaterasu, making it the world's most ancient monarchy. However, after the surrender of Japan to the Allies in 1945, the emperor was no longer considered the head of state and became a constitutional monarch with very limited powers. He is still described, though, as Tenno, or "heavenly sovereign," and is regarded as a symbol of national unity, as demonstrated by his television address to the Japanese people after the earthquake and tsunami of March 2011, which damaged a nuclear power station. Also, December 23, the emperor's birthday, is a public holiday in Japan, as is April 29, the former emperor's birthday. The emperor of Japan led the memorial service in Japan to mark the 1-year anniversary of the earthquake and Fukushima disaster, all proving that the emperor is far more than just a figurehead.

Jimmu Tenno had an ally, a three-footed crow by the name of Yatagarasu, who was sent by Amaterasu to assist with a military campaign in the East. (The crow had already proved himself to be invaluable by suffocating a creature that tried to swallow the Sun.) The crow helped Tenno by discovering pathways in a seemingly impassable land. The crow was partially responsible for the later unity of Japan, and the bird is still worshipped in some shrines. The crow has been included within the emblem of the Japan Football Association, a link between Amaterasu, astro-mythology and the present day in Japan.

The contribution of Arabic astronomers and mathematicians to our knowledge base cannot be ignored. Such astronomers as al-Khwarizmi and Habash al-Hasib thrived under the patronage and encouragement of Caliph al-Mamun (813–833). Like his father before him, al-Mamum recognized the importance of learning, astronomy and mathematics. He founded the 'House of Wisdom' in Baghdad, a scientific institution that included the first large, historically important library since that of Alexander the Great's centuries before. Caliph al-Mamun built two observatories in Baghdad, the capital city of the Islamic Empire and also at Tadmor. The caliph's investments bore fruit, with his astronomers being credited with the use of what we now call Arabic numbers, the discovery of the altitude of the Sun, measurements of the size of Earth and tables of motions of the then-known planets.

Perhaps the most influential astronomer/mathematician was known as al-Khwarizmi, from whose birthplace name the terms 'algorism' and 'algorithm' derive. He can be regarded as the founder of algebra as a separate discipline from geometry. He wrote the *Sindhind zij,* which included how to calculate eclipses, spheres and planetary positions, among other things. He also developed the geography of Claudius Ptolemy.

Certain colors are associated with the Sun. In the Bible, white is the color of light, both physical and spiritual. The color white for the Aztecs symbolizes the dying Sun, or night. The color yellow was the imperial color in China, sacred to the emperor and connecting him to the yellow Earth.

Yellow is a lucky color in China. However, yellow can have rather negative connotations. To call somebody 'yellow' is to call him or her a coward. Yellow can obviously symbolize the light of the Sun, but can also represent treachery and secrecy. In Elizabethan England's color symbolism, true yellow was equated with joy, while lemon-yellow represented jealousy. Orange, both the color and the fruit, can represent the Sun. The fruit looks like the Sun, being a bright orange sphere (Fig. 2.2). Oranges are a lucky symbol for the Chinese and Japanese. The color can also represent fire and flames as well as the Sun. In Elizabethan England's color symbolism, orange is equated with spite. The term 'black Sun' is used to describe a formation of millions of starlings that gather in Denmark for a few weeks in the year, making incredible synchronized shapes in the sky on their migration to and from Europe and Scandinavia.

Fig. 2.2 Oranges have been a symbol of the Sun. (Illustration by the author)

Fig. 2.3 A Chough, associated with King Arthur in mythology. (Illustration by the author)

The Sun is associated with some birds and animals. The chough is protected in Cornwall because it was believed that the soul of King Arthur, Lord of the Light and a solar warrior, was turned into one (Fig. 2.3).

The cuckoo represents the Sun in Hindu mythology. There is a cuckoo festival every year in West Yorkshire in England (Fig. 2.4). People were aware that the arrival of the cuckoo always brought fine weather. They did not want the sunshine of spring to stop, so they built a wall surrounding the cuckoo's nesting place (the idea being that if they could keep the cuckoo, they could keep the good weather). The wall that they built was not high enough, and the bird flew away. This story has now become a weekend festival in April with events such as maypole dancing, itself connected with fertility and the Sun, the participants dancing around the pole re-enacting Earth's journey around the Sun.

Fig. 2.4 Cuckoos are harbingers of spring in England and represented the Sun in Hinduism. (Illustration by the author)

Fig 2.5 The colors in a pheasant's feathers associated it with the Sun. (Illustration by the author)

For the Chinese, the swan is a solar bird. The tears of the Egyptian god, Ra, when they fell to Earth, transformed into working bees, and for the Chinese, the three-legged raven lives in the Sun, the legs representing the three perceived solar phases: rising, at the pinnacle of the heavens and setting.

The pheasant is associated with the Sun by its beauty and color (Fig. 2.5).

The raven is a messenger of the Sun god and sacred to both Helios and Apollo, and in Russian legend, the quail represents the Sun with the hare as the Moon (Fig. 2.6). The goose is said to follow the Sun on migrations, and the Michaelmas

Fig. 2.6 In Russian myth, the quail is associated with the Sun. (Illustration by the author)

Fig. 2.7 Geese represent a solstice turning point in holiday celebrations. (Illustration by the author)

and Christmas geese represent the waning and then rising power of the Sun (Fig. 2.7).

The eagle, according to Mexican thought, represented the Sun, which needed human hearts and blood to continue moving through the sky. The eagle was also thought to be able to fly up to the Sun and gaze on it without blinking. It was also believed to be able to renew its feathers by flying all the way to the Sun and then dropping into the sea, symbolic of resurrection, similar to the Phoenix, endlessly renewing itself in fire. To the Sioux tribes, the eagle's feathers are seen as the Sun's rays. The cockerel, sacred to Apollo, crows at dawn for the victory of the light over

Fig. 2.8 Cockerels or roosters were a sacred sign of Sun god Apollo. (Illustration by the author)

Fig 2.9 A stag's antlers represent rays of light from the Sun. (Illustration by the author)

Fig. 2.10 Hedgehogs have also been linked to the Sun in legend. (Illustration by the author)

darkness (Fig. 2.8). For the Egyptians there was the Benu bird, which looks like a large heron. This bird was later known as the phoenix to the Greeks, as it was a relative of the Sun god.

The stag's antlers are an obvious symbol of light rays and fire (Fig. 2.9), and the hedgehog is regarded as Sun-like probably because of its radiating prickles (Fig. 2.10).

The dolphin is the chosen form of the Sun and Apollo when he carried people from Crete to Delphi, where they built the famous temple to him. The leopard is associated with the light of the morning Sun in African mythology, and the Aztec dog god Xoltl led the Sun though the underworld, where they were both reborn at dawn. The lion, like the eagle, is supposed to be able to gaze directly at the Sun. The lion killing the boar represents the Sun killing the boar of winter, and the solar hero slaying the lion is the Sun god changing the heat of the Sun at noon. For the Chinese, the lion with a ball represents either the Sun or the Cosmic Egg.

Some Christians believed that lion cubs were born dead and life was breathed into them by the male, symbolic of resurrection. For the Egyptians, the lion with the solar disk represents the Sun god, Ra, and for ancient Greeks, the hero Heracles wrestled with a lion, symbolic of the solar hero overcoming death. The yellow tiger is the center, the Sun and the ruler for the Japanese, and the mythical gryphon is symbolic of the light of the dawn turning to gold. For the Druids, the bull is the Sun and the cow, the Earth, and for the ancient Greeks, the bull is symbolic of Zeus as sky god. The celestial bull ploughs the great furrow in the sky for the Sumero-Semitics. The spider in the center of its web can be symbolic of the Sun and its rays.

Other objects including flowers and even food can be associated with the Sun. Flowers, because of their Sun-like center and the fact that they always face the Sun, are associated with it. The sunflower is an obvious example, as it resembles the Sun itself with its golden petals as rays. The cow is associated with the Sun for ancient Egyptians. Their goddess Hathor wears cow horns on her head with a solar disk in between them. Golden roses are symbolic of perfection, and the white rose is the 'flower of light.' The Scarlet Pimpernel or Shepherd's Sundial or Poor Man's

Weatherglass is associated with the Sun, as the flower only opens up on a bright, sunny day. The lotus is a product of the Sun, flowers only in the Sun and is regarded as the light of heaven. Lights on Christmas trees are supposed to be imitating the Sun, Moon and stars shining in the Cosmic Tree. For Mexicans, their Cosmic Tree is the Agave, a cactus with the falcon, representing the Sun's power.

Some animate and inanimate objects can represent the Sun. The wheel and the umbrella share the Sun's symbolism in the center with the spokes, the Sun's rays as the spokes. The heart is the center of life, with veins like the Sun's rays. Arrows are obviously associated with hunting and the Sun's rays, and buttons on Chinese ceremonial robes are symbolic of the Sun. The points on a crown are supposed to represent sunbeams, with the people who wear the crown, usually royalty or rulers, as personifications of a heliocentric cosmology (Fig. 2.11).

Within the Star of David, the main symbol of Judaism, the triangles symbolize the Sun, fire and male energy interconnecting with the Moon, water and feminine energy, while the swastika, before its adoption by the National Socialist or Nazi party in Germany, is an ancient symbol of the Sun and the wheel of birth and rebirth. The golden bough is the connection between this world and the next, while the spiral staircase shows the movement of the Sun. Even food can recall the Sun, especially anything spherical or anything that can represent the rays. Saffron, being yellow, is the herb of the Sun, used in, among many other things, saffron buns.

The Sun is part of our everyday language. We use words and phrases such as 'radiant,' 'sunny outlook,' 'sunny disposition,' 'sunny personality' and 'little ray of sunshine' all the time to describe positive people. People make major decisions on house purchases according to the 'south-facing' direction of the Sun or having a 'suntrap garden.' Some creatures are named after the Sun. A small black Asian bear is called a sunbear, a tiny, brightly feathered bird resembling a hummingbird is called a sunbird (Fig. 2.12). A red starfish with many rays is called a sunstar, and a sunfish is an almost spherically shaped fish (Fig. 2.13). A solar month is one-twelfth of the solar year, a solar wind is a continuous flow of charged particles from the Sun, the solar plexus (a complex of radiating nerves at the pit of the stomach) is the center of the human body and a solar myth is like the myths in the chapter, tales that involve, symbolize or attempt to explain the Sun or solar phenomena.

The Sun certainly affects the weather here on Earth. Scientists have concluded that recent harsh winters have been due to a dip in the Sun's radiation levels. This also explains cold periods such as the Little Ice Age, which occurred between the thirteenth to the middle of the nineteenth century, when winter was about 2 degrees

Fig. 2.12   A sunbird (Illustration by the author)

Fig. 2.13   The origins of the sunfish's name are obvious given its circular shape. (Illustration by the author)

cooler and the River Thames in London froze over. The Sun is also indirectly responsible for the terrible weather in Britain in 2012. April and June 2012 were both the wettest for more than a century, and June was also one of the most overcast, with a mere 118 h of Sun. This is due to the jet stream, a fast-moving band of air that flows above the atmosphere.

Fig. 2.14 The Sun above is an unavoidable influence on all cultures. (Illustration by the author)

The jet stream redistributes heat and energy. Jet streams are caused by the planet turning on its axis and by the atmosphere heating up, in the case of Earth by radiation from our parent star, the Sun. The Sun's influence obviously affects crops and food supply on a global scale. While on a more local level, 2010 was the coldest winter for 23 years in the southwest of England. This made grass taste bitter, as it had received virtually no Sun, so lambs did not like the taste of it. Farmers had to feed them other things.

People are likely to have been worshipping the Sun for thousands of years in some way. When people first walked the Earth 200,000 years ago, they would have looked up at the sky and would have had to have noticed the Sun, Moon and stars. The Sun, being visible in daylight, would have been associated with lighter skies, clear visibility and feeling warmer in some way.

However in Neolithic times (between 4,000 and 6,000 years ago), people have been able to forge a link between themselves and the celestial bodies seen in the sky. This eventually resulted in the construction of stone monuments in Britain and Ireland, such as the obvious example of Stonehenge and others like the Stones of Stennes, constructed around 3000 B.C. which had some sort of cosmological alignment. This was continued with the construction of pyramids. At around 1500 B.C., the Sun Pyramid was built in Mexico, and sundials were being used in Egypt.

More recently, buildings for worship of the Sun were constructed, including the Deyr El Zaffron, or Saffron Church, near Mardin in eastern Turkey. This is known to have been a religious site for 4,000 years and has an underground chamber that was reputedly a site of Sun worship. There is no mortar between the stones on part of the building, just nothingness. Saffron, as said earlier, can be associated with the Sun because of its color, but also associated with love and magic.

Today, our scientific understanding of the Sun is greater than that of our ancestors. The Sun and Moon are an intrinsic part of the Celtic year, and Druids and New Age believers still flock to Stonehenge at the solstices. Despite the rise in skin cancer, some people still desire to lie in the Sun to acquire deeper-colored skin, often referred to as 'Sun worshipping.'

Astro-mythology and its related symbolism illustrate just how much the Sun exerts its influence on us all (Fig. 2.14).

A solar eclipse occurs when a new Moon passes directly in front of the Sun, appearing to block it from Earth. This has been regarded as an eerie, frightening occurrence throughout history, inciting doom-laden stories. There is something deeply unsettling about the disappearance of a heavenly body as familiar as the Sun, albeit briefly. It is unsurprising that an eclipse is regarded as rather auspicious, when unknowing ancient peoples without astronomical knowledge became absolutely terrified thinking about the end of the world. A solar eclipse is only seen over a small area because the shadow of the Moon is small when extending to Earth.

To the Hindus, an eclipse is caused by the bodiless demon Rahu swallowing the Sun or Moon, which goes down through his neck and then back up into the sky. Whereas, to some Muslims and Buddhists, the eclipse of the Sun is sometimes connected with its death, when the Sun was thought to have been devoured by a monster. In China, solar eclipses were seen as the somewhat unnatural dominance of the yin, the Moon, feminine principle over the yang, the Sun, the masculine.

The eclipse is a short demise of the light, a triumph of darkness implying a universal confusion and lack of balance. When a solar eclipse occurs, the Sun and Moon work together, with neither dominating the sky. In a Chinese myth, it is believed that the Sun or Moon has been consumed by a cosmic toad, whereas in Central America, it has been eaten by either a serpent or a jaguar. For inhabitants of the east of India, it is a black griffin, whereas the Persians believed that the heavenly body had been eaten by a dragon.

The ancient Egyptians believed that the serpent Apep swallowed the Sun god, Ra, who traversed across the sky in his boat during an eclipse.

According to superstition, the Sun is supposed to shine brighter on the virtuous but will hide its lovely shining face from us, such as during an eclipse, if something disastrous is going to happen. The people of Peru believed that an eclipse was a bad omen that forewarned of the Spanish annihilation of the Incan Empire.

There is an Inuit myth about an eclipse where a youth called Alinnaq secretly lusted for his own sister. One dark night, he could control his lust no longer, got into his sister's bed and had sex with her. In the morning, the sister was shocked and disgusted that she had had sex with her brother, so she cut off her breast and expected Alinnaq to eat it before running off into the dark night with a torch. Alinnaq tried to find her with a torch, but he stumbled in the snow, which almost annihilated the light of the torch. They both ascended high up into the sky, where the girl became the Sun and Alinnaq, the Moon. He continues to chase his sister, and, during an eclipse, the Inuit think that he has captured her.

Chapter 3

A Crippled Craftsman, a Divine Messenger, and the Goddess of Love

Vulcan

Vulcan is a completely fictitious planet fabricated by scientists in an attempt to explain the planet Mercury's strange elongated elliptical orbit approximately one and a half centuries ago.

Vulcan, who was Hephaestus to the ancient Greeks, was the god of fire and a heavenly metal smith. Some myths tell that he was a child of Hera's, without any father. Apparently, Hephaestus was weak and puny when born, so Hera decided to abandon him and threw him from the top of Mount Olympus. The helpless baby luckily fell into the sea, where the sea goddesses Thetis and Eurynome found him. They bought him up under the sea, where he set up his first metallurgical workshop. There, he made objects for them, including a lance and the three-legged cauldron. All of this ended when, 9 years later, Hera discovered that her rejected son was not only alive but had a talent for creating exquisite or useful metal objects and brought him back to Mount Olympus, providing a more sophisticated metallurgical workshop and forge.

There is a variation of the myth of his birth that claims Zeus as his father by seducing Hera disguised as a cuckoo.

Hephaestus was supposed to have been unattractive, with a powerful upper body from his metalworking activities and a volatile nature. He was not always lame.

© Springer Science+Business Media New York 2015
R. Alexander, *Myths, Symbols and Legends of Solar System Bodies*, The Patrick Moore Practical Astronomy Series, DOI 10.1007/978-1-4614-7067-0_3

His lameness came from Zeus's anger at Hephaestus's rather misplaced loyalty towards his mother, Hera. When Hera dared to rebel against him, Zeus suspended her by her wrists from Mount Olympus. Hephaestus either foolishly or courageously rebuked Zeus. He was flung from Mount Olympus once again. This time, he was not so lucky, falling on the island of Lemnos and breaking both legs when he fell. Eventually, he was forgiven and allowed back up to Olympus, but his broken legs never healed properly.

In some myths, Hephaestus was married to Charis, one of the Graces, her name meaning "to rejoice." Charis was the daughter of Zeus and Eurynome. In another myth, Hera arranged for him to marry the beautiful goddess of love, Aphrodite. Aphrodite, however, was not happy about marrying an unattractive lame god and was unfaithful. Aphrodite's infidelity with Ares, the passionate god of war, was rather blatant and embarrassing for Hephaestus. He decided to teach them both a lesson. He made a strong fine mesh net from bronze and hung it above Aphrodite's bed. He waited until Ares and Aphrodite were having sex and then released the bronze snare, catching them red-handed. As if that were not humiliating enough, he got all the other gods to come and laugh at them, naked and deeply embarrassed in their bronzed cage. Eventually Zeus made him release them.

Hephaestus made many of the amazing and enchanted metal objects that appear in Greek mythology. He crafted the invincible sword that Mercury gave to Perseus and the sickle with which Medusa was decapitated, Mercury's infamous winged helmet and probably Mercury's gleaming golden flying sandals as well as Hades' wonderful helmet of invisibility. He also created the shield and armor worn by Achilles in the Trojan War. Hephaestus's metallurgical workshop came in useful during the Revolt of the Giants, when he flung a container of red-hot metal at the giant Mimas.

Hephaestus, allegedly, tried to rape Athena, virgin goddess of wisdom and war. To avoid losing her cherished virginity, she suddenly vanished. That disappearing act left Hephaestus's sperm to fall to the ground, from which grew the slithering serpent Erichthonius.

Perhaps Hephaestus's greatest creative achievement was the woman Pandora. She was very beautiful, enough for the Titan god Epimetheus to be seduced by her looks, which hid the fact that she was also foolish and indolent. She carried a box with her, safely enclosing potential tortures and extreme annoyances for humans that included Age, Work and Pain. Pandora stupidly opened the box, inflicting its contents on the world of unfortunate mortals, with only Hope remaining, personified by a sprightly crow, holding onto the lid.

It seems appropriate for this fictitious planet to have been called Vulcan, as he was the fire god and god of volcanoes, bearing in mind its imagined closeness to the Sun.

Vulcan is also the name of a type of nuclear bomber aircraft and the race of the character of Spock in the *Star Trek* series and films.

Mercury

Mercury is the smallest of the inner planets, a tiny rocky sphere a mere 4,878 km in diameter. Because of its relatively diminutive size and extreme heat, being the nearest planet to the Sun, it was not able to retain its atmosphere. An atmosphere acts as a protective shield, dispersing attackers such as meteors and comets and fragmenting them, which explains why Mercury was struck repeatedly in its earlier life, creating the cratered and textured world that we know today. This lack of a protective shield for Mercury has also caused its huge differences in temperature. Mercury has no known moons, and its main physical feature is the Caloris Basin, a massive 1,300-km scar of a particularly brutal attack. Mercury has areas of dark, igneous rock from volcanic activity. It is conjectured that its disproportionally large iron core is due to an impact with another small planet, with the attack itself causing some of Mercury's crust and mantle to become casualties. It is the most dense of the planets in our Solar System (Fig. 3.1).

This tiny, hot planet careers around the Sun, the fastest-moving of any of the planets in our Solar System, and so is appropriately called Mercury, after the

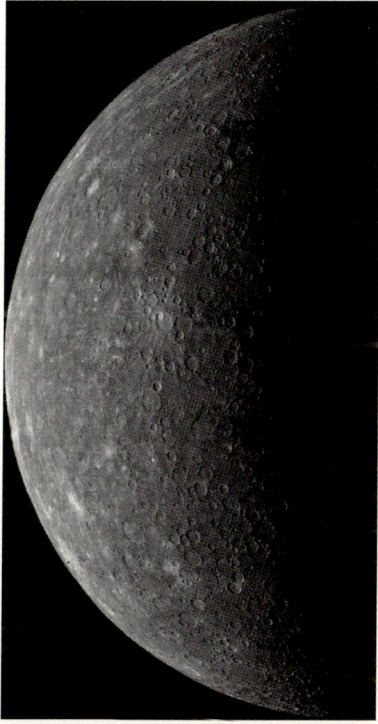

Fig. 3.1 A view of Mercury (Courtesy of NASA)

messenger of the gods, who could fly at incredible speeds and wearing his helmet and sandals that are equipped with wings. NASA's probe that was sent to the planet Mercury was appropriately named 'Messenger.' The United States also launched Project Mercury, with the aim of placing an astronaut inside a capsule orbiting Earth.

Thanks to NASA's Messenger probe, the most mysterious planet in our Solar System has given up some of its secrets, but far from providing answers, these secrets just leave us asking even more questions. It has been conjectured for around 20 years that Mercury has water on its surface, and because of Messenger's data, we now know that the planet has billions of tons of water-ice at the north pole. The most recent information at the time of this writing is the discovery of certain chemicals that could possibly show that Mercury could have formed further away from the Sun at a later time.

Mercury features in the Periodic Table with the atomic number 80 and the symbol 'Hg.' Mercury itself is a silvery liquid metal used in thermometers and barometers. The phrase 'the Mercury is rising' can be used metaphorically to signify that things will get better. Mercury became the god of alchemy, as he bears the same name as the liquid metal, which was also known as 'quicksilver' at a later period in history. Mercury silver is also used on mirrors and on glass for interior design items such as lamp bases.

In mythology, Hermes/Mercury was a son of Zeus and Maia (one of the Pleiades). Though he was mostly known as the messenger of the gods, he was also a god of persuasive speech and languages, theft and deception. He might be seen to be a somewhat benign, trickster character, but he, like most of the gods, had questionable morality and a dark side. He was more than just a messenger in the myth of Jupiter and Io. Jupiter's jealous wife, Hera/Juno, had suspected Jupiter of having sex with the beautiful Io. She turned Io into a beautiful white cow, guarded by her faithful servant Argus, known as the watchman with a 100 eyes. Jupiter sent Mercury to make Argus sleep, a difficult task with someone who had so many eyes. Mercury disguised himself as a shepherd. Argus had a reputation for being boring, and so was unused to company. He was delighted that Mercury wanted to talk to him. Mercury tried everything to make him sleep, resorting to telling endless, excruciatingly boring stories until 98 of his eyes were tightly closed. Then, he shook a magic bunch of poppies over the unfortunate Argus. The sinister magic poppies caused his final open pair of eyes to feel heavy and full of sleep, so they closed. He was finally asleep. Then, Mercury murdered the faithful watchman and led Io the cow back to Jupiter, who changed her back to her original form. Juno placed Argus's 100 eyes in the peacock's tail, in remembrance. It could be argued that Mercury was only acting on Jupiter's orders, but he was still capable of murdering someone whose worst crime was to be boring!

Mercury was also quite happy to sleep with the promiscuous and married goddess of love, Aphrodite/Venus. They became the parents of Hermaphroditos, a female boy from where the word 'hermaphrodite' originates, referring to a person or creature combining both sexes. They were also parents to Priapus, the god of reproduction, a tiny god with an oversized penis from whom we derive the word priapic, meaning "phallic."

Mercury fell in love with a most enchanting virgin, Herse (a moon of Jupiter). He was completely smitten by her beauty and raped her, producing two sons, Cephalus and Ceryx.

It was the duty of Hermes/Mercury to lead the souls of the dead to the Underworld to await their judgment. He also ensured that Persephone was dispatched from the Underworld to her mother, Ceres, on Earth each year.

Mercury also had the dubious honor of offering the beautiful Pandora to Epimetheus as a bride, who was accepted, and then delivered a message to Calypso, ordering her to release her captive, Odysseus.

Mercury could also be helpful. He gave the famous Greek hero Perseus help by telling him where to find the Graeae, who could, in turn, assist him in exactly how to kill the terrifying, stone-metamorphosing Medusa. Mercury provided an indestructible sword for Perseus, who reputedly decapitated Medusa with a sickle, also a gift from Mercury. In the Revolt of the Giants (supposedly the retaliation by the 24 giants on the Olympian gods, as Zeus had imprisoned their brothers, some of the Titan gods, in Tartarus), Mercury, wearing Hades' invisibility helmet, killed Hippolytus. He was also largely responsible for enabling Zeus to escape the clutches of Typhon. Typhon was a gargantuan and fear-inspiring monstrous creature who had Zeus trapped in its horrific serpentine coils. The Typhon was unable to kill Zeus, as he was immortal, but was able to severely limit him by removing tendons from his hands and feet. The tendons were put in the Corycian Cave, guarded by another horrendous creature, Delphyne. Mercury and Pan worked as a team to restore Zeus's tendons and power, with Pan distracting Delphyne while Mercury recaptured the tendons, returning them to Zeus. Mercury also helped Odysseus, whose men had swallowed a potion disguised within a drink and were consequently transformed into pigs by Circe the witch. Mercury had forewarned Odysseus, so he refused the drink. He also used the herb (called moly) that was also a gift from Mercury to transform back his men.

Mercury and Apollo both lusted after the beautiful virginal young girl named Chione. Mercury was the more impatient of the two immortals. Mercury was consumed by his lust, which had to be satisfied. He couldn't even wait until nighttime. He made her fall asleep with his enchanted wand, and then proceeded to rape the unsuspecting 14-year-old. This rather bewildering union produced a son, Autolycus, an expert in deception, like his immortal father.

Mercury not only saved Zeus literally from the clutches of the terror-striking Typhon but also saved the unborn deity Dionysus from Semele's ashes. The unfortunate Semele had been struck by one of Zeus's claps of thunder. Mercury stitched the embryonic Dionysus into the thigh of Zeus, from which he emerged a few months later.

Mercury acquired his trickster reputation when still a baby. He stole an entire herd of cows from Apollo, whom he tricked by making shoes for each cow from the tree bark tied onto each of the cows' four feet. He had to explain himself to Zeus

Fig. 3.2 Mercury was said to have stolen away a herd of Sun god Apollo's cows. (Illustration by the author)

about this and other things. Zeus was impressed by Mercury's ability to persuade and induce and made him his messenger (Fig. 3.2).

Mercury is usually personified as a young man wearing the infamous gleaming, golden, flying sandals and a helmet with wings attached. He sometimes holds a phallic-looking staff, as he is unsurprisingly associated with fertility. He can be shown holding a lyre, which he is crediting with inventing, or sometimes holding poppies, as he was sometimes known as a god of sleep, of which the poppies are a symbol. Zeus has been known to deliver predictions foretelling future events in dreams sent to Mercury.

Mercury is associated with Wednesday, the French word for this day being *Mercredi,* and his staff is his symbol, with two serpents coiling around it representing healing and peace. He is also associated with the colors silver, purple and gray-blue.

Although Mercury has no moons, it does have many, many craters on its surface. They are named after famous writers, artists and composers, although Gerard Kuiper has a crater named after him here as well as on the Moon. Many artists have a Mercury crater named after them, ranging from the ancient Apollodorus of Damascus (thought to have flourished around the end of the first century to the beginning of the second and believed to have been an impressive architect designing, among other things, a bridge that crossed the Danube, baths and a public market) to Renaissance artists Bernini, Raphael and Michelangelo. In addition, there are craters named after famous Japanese artists such as Andro Hiroshige (born in 1779 and most famous for his *The Great Wave*) and Suzuki Harunobu and more recent artists including Spain's Salvador Dali and Pablo Picasso and France's Paul Cezanne, Claude Monet and Auguste Rodin, the sculptor. Composers included Sergi Rachmaninoff, Frederic Chopin, Wolfgang Amadeus Mozart and Guiseppe Verdi. There are even more writers, playwrights and poets that have craters on Mercury named after them. These include the most obvious—William Shakespeare, Homer, Hesiod, Ovid, Sophocles and Charles Dickens—as well as the Icelandic historian and poet Snorri Sturlusson (1179–1241), whose writing included the *Prose Edda.* Many other writers, composers and artists have craters on Mercury named after them that are not included here.

Venus: The Morning and Evening Star

The planet Venus, like all the planets known to the ancients, was not discovered by an individual. It is visible with the naked eye and appears reddish in color in the night sky. There is evidence that around the same time as Homer wrote the epic poems *The Iliad* and *The Odyssey,* astronomers from China and Babylon had some knowledge of the movements of the classical planets. Because of trade, this knowledge would have reached ancient Greece and could possibly have influenced the writings of Homer in some way.

Venus is a volcanic world hidden beneath a thick cloudy atmosphere. This cloud reflects most of the light of the Sun, which shines on it and which is why it appears so bright. It is seriously hot, hotter than the planet Mercury, and spins in the opposite (retrograde) direction to almost all the other planets.

Venus was also known as Aphrodite, the Morning Star or the Evening Star. As Venus shines the brightest of the classical planets, it is appropriate that it is named after the Roman goddess of love, the most beautiful of all the Olympian goddesses. However, she was claimed as a love goddess by people's and cultures long before the ancient Greeks or Romans. Venus has often been mistaken for UFOs in more recent years.

Venus takes 225 days to travel around the Sun and has no moons. It is extremely hot, as its atmosphere is amazingly thick. It is also known as 'Earth's twin,' as it has a similar size and mass to our own planet. It was and still is a volcanic world, and scientists have discovered that it resurfaced itself around 300 million years ago, comparatively recently in the planet's violent history. There is no evidence of water on Venus.

Aphrodite was the Greek goddess of sexual love, lust, beauty and fertility, so it should come as no surprise to discover that she was rather promiscuous. Her own birth is the subject of more than one myth. Some myths proclaim that she was born from the sliced-off penis of Uranus or the foam that congregated around his severed genitals. Her name is derived from the Greek *aphros,* meaning "foam," as she came out of the sea. According to legend, if you swim all the way around the Rock of Aphrodite in Cyprus/Crete, then eternal beauty will be yours. Another myth of Aphrodite's birth is that Zeus sired her by Dione, a sea nymph or a goddess of the oak tree. (Dione is the name of one of Saturn's moons.) Aphrodite is often featured with doves and sparrows, both associated with lasciviousness.

Fig. 3.3 The beautiful Aphrodite, or Venus, stars in many Greek and Roman myths. (Illustration by the author)

Aphrodite was married to Hephaestus/Vulcan, the lame smith-god, regarded as somewhat unattractive. Unsurprisingly, she was frequently unfaithful. Incredibly, Zeus himself, a serial adulterer, did not sleep with the most beautiful of the female deities. He did, however, present her with a gift of an enchanted belt, which made everyone fall in love with whoever was wearing it—not that a goddess of Aphrodite's awesome beauty needed it! She had a passionate affair with Ares/Mars, the god of war. This union produced his three children: Phobos, Deimos and Harmonia. The Sun told Hephaestus, her husband, of this affair. She retaliated by causing him to fall madly in love with the beautiful Leucothoe, whom he raped. The unlucky Leucothoe was buried alive by her disgusted father. The heartbroken Sun could not save her, but was able to change her into a frankincense shrub. Aphrodite was humiliated by her husband, who made a bronze net that he lowered over her and Ares when they were in bed together. They were then laughed at by the other gods and goddesses. However, this punishment by Hepheastus backfired, as the sight of the beautiful Aphrodite naked caused Poseidon/Neptune and Hermes/Mercury to fall in love with her or lust after her. Aphrodite truly did not need the enchanted belt, which she certainly was not wearing while caught in the net. Predictably, she had affairs with the gods Poseidon and Hermes (Fig. 3.3).

Fig 3.4 Anemones represent Adonis, changed into a flower after he was fought over by the gods. (Illustration by the author)

Aphrodite seemed to enjoy causing heartache to others besides the Sun. She made Eros/Cupid fire an arrow that wounded Hades/Pluto in the heart. This led him to fall in love with and kidnap Persephone.

Aphrodite/Venus fell in love with the handsome young Adonis, who was also loved by Hades/Pluto's wife, Persephone/Proserpina, queen of the Underworld. Venus, according to Ovid, even removed herself from the sky, as she was so much in love with Adonis. Despite warning her lover to avoid boars and lions, the jealous Ares/Mars caused a boar to kill him. (Ares still loved Aphrodite, even after the bronze net fiasco.) The death of Adonis caused a huge fight between Persephone and Aphrodite, which was ended by Zeus himself. He decided that Adonis should spend 4 months of the year each with himself, then Persephone, then Aphrodite. In a variation of this myth, Aphrodite changed Adonis into an anemone or wind flower. The Greek word *anemos* means "wind" (Fig. 3.4).

Zeus made Aphrodite fall in love with Anchises, king of the Dardanians. They produced a child, Aeneas, together. Anchises was warned that he must keep the fact that they had a child together a secret. Predictably, Anchises failed to keep his mouth shut and was blinded. Aeneas was important in the legendary Trojan War, and because of him, some Trojans escaped. However, Aphrodite was one of the catalysts of that epic war. Paris, a Trojan prince, was given the unenviable task of choosing the most beautiful of three goddesses: Aphrodite, Hera and Athena. They all offered bribes, but Aphrodite's promise of his having the most beautiful woman in the world as his wife was the most tempting. Paris allowed his emotions to rule and chose Aphrodite's bribe. The beautiful woman was Helen of Sparta, better known as Helen of Troy, who is blamed for the Trojan War.

The sexual Venus was supposed to have hated the innocent appearance of the lily and added to it a very phallic-looking pistol, according to legend.

Fig. 3.5 The brilliant colored quetzal is linked to Aztec deities. (Illustration by the author)

For the Aztecs, Quetzalcoatl was a major deity. The name means "feathered snake or serpent," (*coatl* meaning "snake"). The god Quetzalcoatl appeared as a snake covered with the iridescent green feathers of the beautiful Quetzal bird. Some believe that Quetzalcoatl was reincarnated as the planet Venus. According to Nahua myth, Quetzalcoatl, while passing in the sky from the west to the east as the planet Venus, came across Mictlantecuhtli, god of the dead and the Underworld. Quetzalcoatl asked this fearsome god for his dead father's special bones that man was to be created from. Quetzalcoatl tricked Mictlantecuhtli and claimed that he had left the bones in the Underworld. He returned supposedly to retrieve the bones but was really holding them close to him, keeping the male and the female bones separate. He gathered them up. The bones were ground up by the goddess Cihuacoatl. She put the pulverized bones into a jade bowl, where Quetzalcoatl pierced his penis so he could drip blood into the bowl from his fertility organ. Then humans were created from this bone-meal. The quetzal is, arguably, a main contender for the title of the world's most beautiful bird and has become the national bird of Guatemala. The quetzal is emblazoned on postage stamps and coins, and the national currency is not a peso or a dollar but a quetzal, an example of how astro-mythology is today an integral part of the human psyche (Fig. 3.5).

The quetzal bird was worshipped by the Aztecs and Mayans and was their god of the air. It is believed that if it is caught and caged, it will die of a broken heart. It is a very delicate bird, too fragile to make stuffing it effective, so it is unlikely to be found encased in museums. It has become symbolic of freedom. Quetzalcoatl, when shown as a mortal man, wears a red mask with a beak and body paint. Sacrifices were made to the deity Quetzalcoatl, blood from the human tongue and the penis and tiny flying creatures such as butterflies and hummingbirds.

Venus, known as both the Morning and Evening Star, is linked it to the Star Boy legend of the Plains tribes. An old woman called Old-Woman-Night adopted Star Boy. He became a hero and a vanquisher of monsters. At the climax of his adventures, a snake slid into his body, remaining there until the boy's demise, even when all that was left of the boy was his bony frame. The snake prevented the boy from being resurrected, and he transcended upwards into the sky, becoming the Morning Star.

The Zuni people have a legend explaining the Morning and Evening stars. They believed that a monster called Cloud-swallower, as the name suggests, swallowed the clouds, which in turn caused a great shortage of rain. They shot him, as they needed rain to fall. The various bits of his body transcended up into the sky. His heart became the Morning Star and his liver became the Evening Star.

The Morning Star features in the Bible. Jesus described himself as, among other things, the 'bright morning star' in the Book of Revelation.

Venus is associated with the metal copper and the weekday Friday, *Vendredi* in French. Venus is personified by a beautiful woman and is associated with the stunning but somewhat fragile quetzal bird. Another bird, the swallow, is also sacred to the goddess. A ribbon-like jellyfish marine creature is called Venus's girdle, and from Aphrodite we get the word 'aphrodisiac,' referring to something that is supposed to arouse sexual desire, such as oysters or strawberries. The five-pointed star, the Star of Bethlehem, represents the eastern star or Venus as the Morning Star. The eight-pointed star symbolizes Venus's equivalent goddess, Ishtar, to the Assyrians and Babylonians. Ishtar was goddess of sexual love and war. This star also represented Venus the planet. Venus is also symbolized by a glyph, supposedly a mirror or a necklace. This is a circle with a cross on the bottom, like a charm on a necklace.

It is appropriate that, with Venus being the only planet in our Solar System that is named after a female deity, that all of her craters should be named after women. There are many, many craters on the surface of Venus. In fact, there so many craters that it was hard to find enough famous names for them all, which is why the largest group of craters are merely given a female first name. The long list includes the names Amelia, Bernice Clio, Caroline, Deborah, Denise, Dheepa, Elizabeth, Emma, Florence, Fatima, Grace, Gudrun, Hannah, Heather, Hiromi, Irina, Jane, Juanita, Judith, Karen, Katusha, Liv, Margarita, Miriam, Noriko, Olga, Pasha, Rachel, Rhoda, Ruth, Sarah, Shakira, Sophia, Tako, Tamara, Ulla, Vanessa, Wazata, Xenia, Yoko, Zeinab and Zoya.

Some famous queens are honored by named Venusian craters. Two of the wives of the most famous English king, Henry VIII, are represented: Anne Boleyn (1501–1536), who was queen of England between 1533 and 1536, and Jane Seymour (1509–37), the second and third of his wives. Henry had genuinely loved Anne at first. She was blatantly pregnant at court before Henry married her. Anne's main problem was the same problem Catherine of Aragon had before her—a failure to produce a legitimate male heir. Anne did give birth to the future Queen Elizabeth I but was publicly beheaded in 1536. Henry VIII became head of the Church of England when he was unable to obtain permission from the Pope to annul his marriage to Catherine, and he publicly denounced Catholicism. Jane Seymour did not fare well either. She was not actually crowned queen, as she died soon after giving birth to Henry's longed-for son, Edward. The Grey crater is named after Lady Jane Grey (1537–1554), who ruled England for just 9 days in July 1553. She was executed in 1554. Mary, Queen of Scots (1542–1587), is the royal whom the Stuart crater is named after. The Isabella crater is named after Isabella of Castille in Spain. She bankrolled the travels of Christopher Columbus. She was an educated woman and educated all of her children, including the girls. One of her children was Catherine of Aragon (1485–1536), who became the first wife of King Henry VIII of England.

The Bathsheba crater is named after King David's wife in the Bible. She was queen of Israel. King David abused his royal power when he saw Bathsheba bathing naked. He discovered that she was married to a soldier, Uriah, whom he had sent to the front line, where he died. King David then seduced her and married her. She gave him a son, who later became King Solomon, famous for his wisdom, wealth and power.

The Cleopatra crater is named after Cleopatra, the Egyptian queen (69–30 B.C.). She ruled Egypt from 51 to 30 B.C. and was famed for her passionate affairs with the Roman emperor Julius Caesar and Mark Antony. She was also an excellent politician.

The Agrippina is named after Agrippina the Elder (14 B.C. to A.D. 33). She was the mother of the mad emperor Caligula.

There are many craters named after famous writers. The Wollstonecraft crater is named after the British writer Mary Wollstonecraft, the 'grandmother of feminism,' (1759–1797) who was the author of *The Wrongs of Woman*. She died days after giving birth to Mary Shelley, the author of *Frankenstein,* which is actually a feminist story with Dr. Frankenstein refusing to create a female 'monster' as a companion to his lonely ugly male creation. Simone de Beauvoir was a French feminist writer (1908–86), author of, among other works, *The Second Sex*.

The Behn crater is named after the English writer Aphra Behn (1940–1689). Behn was the first professional English female writer. She wrote plays and the novel *Oroonoko*. She was an independent professional woman who enjoyed celebrity status in her time. Agatha Christie (1890–1976) was a prolific English crime novelist most famous for her fictional detectives Miss Marple and Poirot; the Christie crater is named in her honor. The Potter crater is named after the English writer, artist and farmer Beatrix Potter (1866–1943), most famous for writing and illustrating many children's storybooks featuring such characters as Peter Rabbit, Tom Kitten and Jemima Puddle Duck. Potter was a successful farmer and sheep breeder and eventually bought much of the land in the Lake District and left it to the National Trust.

Female artists include the French Impressionist painter Berthe Morisot (1841–1895); the American artist Georgia O'Keeffe (1887–1986), famous for her oversized flower paintings; Frida Kahlo, the Mexican artist (1907–1954), most famous for her dramatic self-portraits; and the outstanding Italian painter Artemisia Gentileschi (1593–1656), the first female painter to be accepted as a member of the Florence Academai di Arte del Disegno. All of these women artists have Venusian craters named after them.

Other craters are named after famous explorers, aviation pioneers, astronomers and scientists that just happen to be female. Many Russian astronomers, such as Evenia Bugoslavskaya (1899–1960), who was a professor of astronomy, have craters named after them, as do pioneering scientists such as the American Sarah Frances Whiting (1847–1927) and Mary Watson Whitney (1847–1921). Sarah Whiting established a physics laboratory at Wellesley College in Massachusetts, the first in the United States for women. Mary Whitney made Vassar College, New York's, research program in astronomy into one of the best in the United States, and in 1889, she was a founding member of the American Astronomical Society.

Aviation pioneers, such as the English Amy Johnson (1903–41) and the American Harriet Quimby (1875–1912), have Venusian craters in their name. Amy Johnson was a pioneer and record breaker. She was the first woman to fly from Britain to Australia, the first person to fly from London to Moscow in a single day in 1931 and in 1932 beat her own husband's speed record for flying from London to Cape Town. She died in a plane crash. The plane was inexplicably off course and crashed into the Thames estuary. Her body was never found. These unusual circumstances fueled rumors. Harriet Quimby was the first American woman to fly across the English Channel.

The Boyd and the Akeley Venusian craters are named after the American explorers Louise Arner Boyd (1887–1972) and Delia 'Mickey' Akeley (1875–1970). Louise Boyd was known as the 'Ice Woman.' She explored and photographed the Arctic and its ocean. She studied glaciers, and the region joined to the De Geer Glacier was later named Louise Boyd Land, and so she has been immortalized both here on Earth and on Venus. She was lucky in that her family wealth gave her more freedom than many other women of her time. However, she chose to spend some of that wealth on exploration, displaying that same pioneering spirit that other explorers had, such as Ernest Shackleton, Sir Robert Falcon Scott and Neil Armstrong. Delia Akeley, along with her husband Carl, traveled by camel. She also wrote books about her amazing adventures.

The Devorguilla Venusian crater is named after Devorguilla of Galloway. She was the wife of John, the fifth baron of Balliol, and when he died she bequeathed a permanent income to the Oxford College that still bears their name, Balliol.

The MacDonald Crater is named after the Scottish heroine, Flora MacDonald (1722–1790). Flora is famous for helping 'Bonnie Prince Charlie,' aka Prince Charles Edward Stuart, to escape from the English. Charles fled to the Scottish Outer Hebrides after being defeated at Culloden in 1746. She allowed him to join her group to travel to the Isle of Skye. Some stories portray Flora and Charles as lovers. Flora was punished by the English for her part in Charles's escape. She was held captive in the Tower of London until eventually being pardoned and released in 1747.

The Godiva Crater is named after the infamous Anglo-Saxon noblewoman, Lady Godiva (1040–1085), who was married to Leofric, Earl of Mercia. Legend has it that Leofric was so tired of his wife's countless appeals to reduce the tax burden of the people of Coventry that he told her that he would reduce their taxes if she rode naked through Coventry's busy market area. He did not expect his wife to actually go through with it and probably regretted ever saying this. Lady Godiva did go through with it. She is depicted in art as having very long hair that covered up most of her body! The men of Coventry stayed inside when Lady Godiva rode through the market, although one man, thereafter known as Peeping Tom, couldn't resist taking a look and was struck by blindness. The people of Coventry were released from all of their tax burdens apart from the tax upon horses. This event, whether real or fictitious, is commemorated each year by an annual procession re-tracing Lady Godiva's ride through the city.

Perhaps there are not enough famous women to name craters after, but that is certainly due at least in part to the lack of opportunities for women until relatively recently. Social, cultural, political and financial constraints have held women back. Louise Boyd had her family fortune behind her, and Delia Akeley was fortunate enough to travel with her husband. In England, until the Married Woman's Property Act of 1882, when a woman got married, her husband owned everything that was previously the woman's, including all of her clothing. This inequality meant that women had to put up with violence, cruelty and infidelity when married. Unmarried women had more power and independence, exemplified by one of Britain's most powerful historical figures, Queen Elizabeth I. If women had had the same opportunities as men, then there would probably be fewer Venusian craters than famous women.

Chapter 4

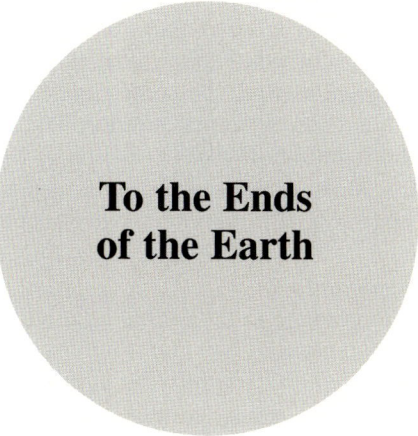

To the Ends
of the Earth

To the ancients, Earth was the most important object in the then-known universe. The ancient Greeks devised a geocentric system, with 27 clear spheres surrounding Earth, each having a celestial object—planets, the Sun, the Moon and stars—joined to it. Round pathways were based on the classical belief that the circle represented perfection, the only possible way that a celestial object would travel. Earth's status was downgraded when the Polish astronomer Nicolaus Copernicus introduced a heliocentric system in writings published in the year of his death and proven by the Italian astronomer Galileo Galilei.

Earth is also known as the Blue Planet or the Goldilocks Planet, the former because it looks predominately blue from space, due to the large expanses of surface liquid water, the latter because conditions are just right for intelligent life to thrive—not too hot like Mercury or Venus, not too cold like the outer planets. At 12,756 km (7,926 miles) in diameter, it is the largest of the four rocky planets. Earth not only has a Goldilocks temperature range but has other Goldilocks conditions, such as easily accessible liquid water, an atmosphere and a reasonably quick rate of spin around a sufficiently aged star.

Earth also has tectonic plate movements that cause earthquakes, tsunamis and volcanic action. The process of plate tectonics causes our planet to change and constantly renew its surface. Certain areas are more active than others. The famous Pacific Ocean Ring of Fire is almost permanently volcanically alive, with the sixth largest earthquake ever recorded hitting Japan in March 2011 followed by after-shocks, a tsunami and an explosion at the Fukushima nuclear plant. Even a country such as England, which is not in an area where tectonic plates meet, can experience mild earthquakes, as fossil plates still cause problems.

© Springer Science+Business Media New York 2015
R. Alexander, *Myths, Symbols and Legends of Solar System Bodies*, The Patrick Moore Practical Astronomy Series, DOI 10.1007/978-1-4614-7067-0_4

It is believed that Earth has abundant internal heat and the capacity to power these tectonic movements. Earth consists of a dense metallic core and a mantle and crust of solid rock. Earth's internal heat dates back 4.5 billion years ago, to when our planet was formed. The impacts on Earth at this time generated heat now trapped inside our planet, half of which comes from radioactive rock and materials. Our crust—inspiring phrases such as 'on solid ground,' 'terra firma,' 'down to earth' and 'grounded'—acts like a thermal duvet or shield, keeping this heat in. Earth is cooling, but this trapped heat wants to escape by way of magma.

Earth's core is vital to life upon our planet, as it created our magnetic field. This not only produces a protective barrier between our planet and the solar wind but acts as a navigational aid, along with the angle of the Sun, the stars and the landscape for the navigation of birds.

Earth is constantly resurfacing itself. Earth's rigid surface is fragmented into tectonic plates. These are constantly moving, driven by internal heat; some plate movements cause earthquakes and tsunamis. All of this makes us realize that our planet is alive and unpredictable, and we are only just beginning to understand how it works. The land mass on our planet 200 million years ago would be unrecognizable to us now. The land sat in one large mass, which we have given the name Pangaea. This land has split apart several times, and parts have even joined together again, creating geological features such as mountain ranges in the process.

In Greek mythology, Poseidon is known as the Earth shaker, and therefore responsible for earthquakes, flooding and tsunamis, among other things. Earthquakes are regarded as a threat to the 'regular,' or cosmic, arrangement, and requires appeasement, including include human sacrifice.

In Norse legend, earthquakes were explained by Loki, the giant trickster and fire god wriggling in extreme physical pain. Loki was blamed for the tragic death of Balder, the favorite son of Odin and Frigg. Frigg had a premonition of Baldur's premature death, and so warned every creature, every plant, not to harm her beloved son. Unfortunately, Frigg omitted telling the mistletoe, who persuaded Hodr, a blind god, into shooting Balder with a mistletoe branch arrow, causing his untimely death. Loki received a very severe punishment for this evil deed. He was fettered under Yggdrasil, the world tree, to rocks and had the poison from a snake constantly running down his face. His dutiful wife, Sigyn, collected this poison in a bowl, but when she momentarily left to empty it, the venom fell again on his face, causing him to twist and squirm.

Earth has huge oceans, including the Arctic, Atlantic and Pacific. In fact, the oceans make up around two-thirds of Earth's surface. The question is, where did this water come from originally? Some scientists believe that the answer lies in comets. The oceans are the final frontiers here on Earth. There is so much of them that we have yet to explore. In the 1960s, oceanauts were doing amazing things. In 1963, an underwater village inside a capsule, called Conshelf 2, was created in a remote place near east Africa. Five oceanauts lived on the seafloor in this capsule for a month at twice the normal atmospheric pressure. The enormous pressure is still our greatest challenge to deep-sea and ocean exploration, which explains why almost 70 % of the deep sea is yet to be explored, as is around 80 % of the Antarctic seabed.

Scientists have recently discovered that ice in the Arctic is getting thinner and therefore weaker and covers a smaller surface area. As white ice reflects the Sun's rays back up into space, and the dark ocean surface absorbs the Sun's rays, the climate is affected. The warmer it gets in the Arctic, the more likely the path and speed of the jet stream (the high altitude wind that guides weather systems will be affected after the Atlantic Ocean). This, in turn, could lead to more storms in Europe.

Coastal locations with potentially dangerous rocks or headlands are the kind of place where there are usually local myths or legends concerning drowning.

At Pendour Cove, Cornwall, there is the legend of the Zennor mermaid. (Zennor is the name of the nearby village.) In the local mythology, there was a young man who went by the name of Matthew Trevella. He was not particularly blessed with good looks, but he did have an amazingly beautiful singing voice. He sang every evening at the local church at Zennor, always singing the final hymn alone, such was the haunting power of his voice. A mermaid who resided nearby in the waters of Pendour Cove was so moved by this charming singing that she felt compelled to find the singer. She attempted to disguise herself by wearing a dress long enough to cover her beautiful tail. She was cautious, at first, and just went to the church to listen to the singing, then went back to her watery residence afterwards. After a while, she plucked up enough courage to remain at the church, and their eyes met. It was love at first gaze for the youthful Matthew. She got up to leave. Matthew stopped her, engaging her in conversation. He learned that she was a mermaid and had to return to the sea or she would surely perish. Matthew was so consumed with love for the intriguing and beautiful creature that he went with her to Pendour Cove, walked into the sea after her, despite the pounding waves. He was never seen after that.

Legend holds that those who linger above Pendour Cove at sunset on a clear summer evening could possibly still hear Matthew's haunting, soulful voice upon the wind.

At Zennor, there is the small church of St. Senera, which originally dates from the twelfth century and is named after a Cornish Christian saint. The burial ground contains many memorials to lost sailors. Inside the church is a medieval bench end carved with an image of a mermaid, believed to be over 600 years old, proving that this mermaid existed at least in the imagination of the person who made the carving, or at least in that of the person who paid the carver so many years ago. The legend, position of the church and graveyard full of memorials to those whose bodies were consumed by the sea serve to remind people of the ruthless power of the sea and the wind.

Stories of maritime tragedies must have influenced the Danish author Hans Christian Anderson's (1805–1875) fairytale, *The Little Mermaid*. This tale also concerns a potential union between a sea creature and a human that does not work out, even though the mermaid gives up her voice in exchange for legs.

Mythology is used to try to provide an explanation for the otherwise inexplicable. An example of this is the famous tourist attraction known as the Giant's Causeway in Northern Ireland. This geological anomaly consists of thousands of hexagonal rocky short pillars, supposedly the handiwork of the legendary Irish giant and hero Finn MacCool. Finn acquired his wisdom from cooking the Salmon

Fig. 4.1 The Salmon of Knowledge caught by Finn MacCool. (Illustration by the author)

of Knowledge and burning his thumb on it. He sucked his thumb to cool it down and, in doing so, acquired the fish's wisdom. He is supposed to have formed a path in the sea from bits of earth so that the Scottish giant Bennadonna could meet up with him to fight.

Most mountains have mythological significance. Mount Olympus is supposedly the legendary home of the Olympian gods ruled by Zeus, and in China, the Tai Shan is supposed to be the mountain of creation where peoples' souls both arrive and exit this earthly realm (Fig. 4.1).

A popular myth of the Himalayas is that of the Yeti or Abominable Snowman. The creature is supposed to be an apelike man, or even ursine in form. The Yeti started to appear in western mythology in the nineteenth century, becoming even more popular in the twentieth century as more people arrived in the area to climb the mountains, especially Mount Everest. The Yeti was believed to have evolved from a hunting deity from before Buddhism took hold.

No real evidence has been found of this creature, just footprints and photographs of footprints, which most scientists are unconvinced by. No photographic or film evidence exists of the elusive being at the time of this writing, but there are a wealth of films and stories.

The Yeti legend is believed to date back hundreds of years to Alexander the Great's time. People back then apparently saw wild-looking, man-like creatures with their feet turned around the wrong way, according to Roman author Pliny the Elder (A.D. 23–79), who wrote *Natural History*, an early encyclopedia. However, he wrote this around 400 years after Alexander's expedition. An interesting fact is that contemporary Tibetans still believe that the Yeti have feet pointing backwards so that they could ascend the mountains very quickly.

During an expedition to Mount Everest in 1923, a British man called Major Alan Cameron claims that he saw very large, dark beings high up on a different part of the mountain. Even the famous Sir Edmund Hillary and Tenzing Norgay, the first men to reach Everest's summit, at one point thought that they saw large footprints.

The popular phrase 'abominable snowman' was not widely used until 1921. In 1937, the British explorer Frank Smythe recounted that his Sherpas and local people had seen a somewhat bestial, man-like creature in the mountains. The abominable snowman, or *rakshas* (demon), myth became popular enough for a wealthy American, Thomas Slick, to fund expeditions to look for the creature. On an expedition in 1959, strange excrement was discovered. When it was analyzed, scientists discovered an as yet unknown parasite, implying that the creature this new parasite had fed on was also new to them. This myth still captures people's imaginations, and it does seem possible that some unknown creature has eluded us in this hostile environment. Recent DNA testing of supposed Yeti hair samples from Bhutan and Ladakh, in the Himalayas, showed that of an ancient polar bear. It is believed that the Yeti is probably a hybrid of a brown bear and a polar bear.

It is 61 years since Mount Everest was first conquered, and over 3,000 people have now climbed this majestic and awe-inspiring mountain. Everest is the tallest mountain above sea level on the planet and is found in the Himalayas on the border of Nepal and China. The Tibetans worship the mountain gods and believe that every tall mountain has a god. It is traditional for Sherpas to write prayers on prayer flags that are set up at the base camp at the beginning of the climbing expedition. The idea is that the wind will take the prayer flags up to the gods. When New Zealand beekeeper, explorer and mountaineer, Sir Edmund Hillary (1919–2008) and Nepalese Sherpa Tensing Norgay (1914–1986) conquered Mount Everest in 1953, they both left small items. Hillary left a crucifix and Sherpa Tensing left sweets for the mountain gods. People from around the Mount Everest region also believed in the Sisters of Longevity, who are supposed to live on Everest's peak. There are five sisters. The most well-known rides a white lion and is in charge of the lifespan and health of people. She carries, among other things, dice made from white shells, suggesting that the fate of man is, to some extent, left to chance or luck from a roll of the dice.

In England, there are collections of stone circles, of which Stonehenge, a World Heritage site, is the most famous. The third largest of these is the lesser-known Stanton Drew stone circle in Somerset. The site itself is believed to date back perhaps as far as 3000 B.C. Legend has it that the stones are guests and musicians from a wedding Satan encouraged to take place on Sunday, the Sabbath. They were punished and made examples of by being turned to stone.

Stones are often used as symbols of the power of the gods. In Greek mythology, Helen of Troy, or Sparta, found two stones that, when rubbed together, produced blood, and Zeus himself was replaced by a stone when Cronos/Saturn swallowed his offspring. Stones were strong, heavy, useful, ancient and permanent. Some stones literally fall from the heavens (in the form of meteorites), so these would have an obvious divine connection. Sacrificial altars were often made of stone, one of the most famous being at Stonehenge in England.

The ancient Greeks and Romans crafted statues of their deities from stone, especially from marble, since statues were literal representations of the gods. In Greek mythology, humans were replenished from stones. The sons of Lycaon angered Zeus to such an extent that he produced a flood, intended to annihilate the human race; this was known as Deucalion's flood. However, the rebellious Titan god, Prometheus, forewarned one of his mortal children, Deucalion, about this, and Deucalion built an ark, thus surviving the flood with his wife, Pryrrha. The flood appeared to cover the whole of the known world and did indeed appear to wipe out the rest of the human race, showing the power of the mighty Zeus. Deucalion and

Pryrrha begged and petitioned Zeus to renew the human race. Themis, the Titan goddess and bringer of justice, instructed them to throw the bones of their mother (meaning Mother Earth's bones, which were stones or rocks) behind them. They did as instructed, and the stones were transformed into people, thus restoring the human race.

The Blarney Stone, a blue stone incorporated into the Blarney Castle building in Cork, Ireland, is also known as the Stone of Eloquence. It is believed that if the stone is kissed, then the gift of eloquence will be bestowed on you. There are many myths about this stone. Some say it was brought to Ireland by Crusaders and that it was formerly the stone which the Biblical King David hid behind when he ran from King Saul or that it was the Hebrew Jacob's pillow. Whether the stone grants eloquence or not, it has been kissed by many pilgrims, film stars and statesmen, including the British former prime minister, Sir Winston Churchill, in 1912, who did become famous for his rousing, inspiring speeches during the Second World War!

The Stone of Destiny, or Stone of Scone, was taken from Scotland in 1296 to England by King Edward I when he conquered Scotland. Despite being a symbol of Scotland, it is incorporated into the coronation chair English monarchs sit on when crowned. It, too, has mythology attached to it, including that it was Jacob's Pillow, just like the Blarney Stone.

Stones have long been associated with fertility. The Truroe Stone in County Galway, Ireland, is a phallic-shaped stone, obviously suggesting fertility. The Swearing Stone is a standing stone with a hole through it and is situated on the grounds of a ninth-century monastery in Castlemot, County Kildare, Ireland. It was probably used to seal contracts by shaking hands through the hole. Even today, stories can be created for memorials and grave stones. On the outskirts of the city of Exeter, England, there is a stone cross at the top of a hill called Little John's Cross Hill. The most likely explanation for the name is that there was a family called Littlejohn, but there is a more romantic and highly unlikely local myth that Little John of Robin Hood's Merry Men came to Exeter. He is supposed to have stood outside Exeter Cathedral and fired an arrow. The stone cross is supposed to mark the spot where the arrow fell. This myth is contained within the name of a nearby hill, Sherwood Rise, after the forest where Robin Hood and his men were supposed to have resided.

In Scotland, there is a famous, deep, dark loch. It is deeper, in fact, than the North Sea and contains black water from the peat of nearby hills. It also has extremely steep sides and a flat bottom. Plants do not grow in this mysterious loch, but plankton has been located 100 ft down.

Loch Ness is famous due to the mythical creature the Loch Ness Monster, or 'Nessie,' apparently first seen in the fifth century and still sighted frequently. In the sixth century, Saint Columba traveled from Ireland to try to save the Scottish people near to Loch Ness who were living in fear of the monstrous creature. The story goes that one of Saint Columba's men swam across the water. The creature made an appearance rather close to the man. Saint Columba is supposed to have made the sign of the cross and the monster apparently vanished. It is supposed to be large—at least 30 ft in length and with a long neck, a head a bit like a horse with lumps on

Fig. 4.2 Major Oak. (Illustration by the author)

its back, perhaps reminiscent of some kind of dinosaur. Scientists are, as yet, unable to prove or disprove the creature's existence, but this mythological monster has captured the imaginations of many people for well over a thousand years.

In England, a 'green and pleasant land' according to the English poet and artist, William Blake, there are many myths relating to Earth. Robin Hood was a legendary folk hero in Nottingham. He and his 'merry men' lived as outlaws in Sherwood Forest and wore Lincoln green, a color associated with renewal, spring and fertility. They famously robbed the rich to give to the poor. They were believed to have gathered under the famous tree, the Major Oak, which is now propped up. Sherwood Forest is all that remains of a much larger forest of Shire Wood. Sherwood Forest is home to hundreds of ancient oak trees that could have been around in Robin Hood's time. The oak tree has become a symbol of England and the oak leaf is a symbol of the National Trust (Fig. 4.2).

The beautiful Georgian city of Bath has history connected to Earth. Legend has it that around 3,000 years ago, Prince Bladud caught a disfiguring skin disease and was banished from the palace. He became a swineherd, but even his pigs caught his disease. He traveled into unknown territory, looking for acorns for the pigs to eat, and crossed the River Avon at a shallow point. (This place is now known as Swineford.) The pigs enjoyed rolling around in the hot mud from near to Bath's springs and were miraculously cured of their affliction. Bladud thought that this might work for him, and he too rolled around in the hot mud. It did work for him, and his disease was cured. He returned to the palace, eventually becoming king. Bladud founded the city of Bath, attributing its seemingly magical curative powers to the Celtic goddess Sul. Almost a thousand years later, the Romans named Bath

Aquae Sulis—the waters of Sul. The popularity of the city was revived by the Georgians in the eighteenth century, where 'taking the waters' became a fashionable pursuit and a mythical cure-all.

Mud appears in other mythologies and legends. It is generally regarded as the primal material that the gods created things from. In Babylonian legend, Aruru was a goddess who formed people as if she were a potter of the New Stone Age, and Prometheus was supposed to have molded humans from clay in Greek myth. This idea of people being formed from the earth is echoed in Islamic myth. It is believed that Allah told his angels or messengers to collect seven different colors of earth. Earth was reluctant to allow this, so the Angel of Death had to steal it. The soil would be returned when the human race were all dead. With this colored soil or earth, Adam and other races that were descended from him were created by Allah.

Marduk, the chief god (and national god) of Babylon went into battle against Kingu, a demon and the child of the goddess of chaos and the primeval saltwater and an array of monstrous creatures. Marduk defeated and killed Kingu and was believed to have combined some earth with the blood of the dead demon Kingu. This earthly mixture was moldable, and Marduk was supposed to have created the first people from this. As for Kingu himself, he was banished by Marduk to the Underworld, Ereshkigal.

The earth beneath our feet or soil or ground is believed to have enchanted, curative properties. In the past, people have buried their live bodies (not completely!) in soil in the belief that it would cure all kinds of illnesses and ailments. Births and deaths were believed to be less painful when they were conducted on the ground, or earth, the potency and strength of the soil believed to have literally rubbed off on them.

In Native American mythology, Earth was created by a raven dropping pebbles to create islands in the ocean.

Earth is usually regarded as feminine, passive and dark, usually a feminine deity in mythology. Earth is the womb from which all life is born but is also the grave to which we all return. "Ashes to ashes, dust to dust." Perhaps this is why, in the English language, Earth is the same word for the ground, the soil, as for the planet itself. Earth is the only planet in our Solar System that is not named after a Greek or Roman deity.

Earth and the ground are often used in phrases and truisms. "It cost the Earth" means it was very expensive, "down to Earth" and "grounded" mean sensible, while "grounded" can also mean confined to the house. The phrases "feet of clay" can refer to a weakness. In the French language, the potato is called *pomme de terre,* or "apple of the ground."

The phrase "down to Earth" could also refer to the character of Antaeus from Greek mythology. Antaeus was the son of Poseidon, god of the oceans, the great Earth shaker, and Mother Earth, or Gaia. He was of giant stature and loved to wrestle. He was rather sadistic, taking great pleasure in forcing men to fight with him until he won, when he killed these unfortunates. He kept the skulls of his victims, like trophies that he displayed on the roof of a temple dedicated to his father. Antaeus' strength seemed eternal, as every time he touched Earth, it was renewed with vigor. He even slept on the ground, obviously further increasing his impressive strength.

Antaeus was a formidable opponent with a terrifying reputation, but Heracles/Hercules seemed undeterred. Before fighting, Antaeus coated his arms and legs with hot sand, giving him even more strength, like an intimidating, pre-fight ritual. Heracles himself was a man of immense strength and courage. After all, he did hold up the heavens temporarily for Atlas and was son of Zeus and a mortal princess, Alcmene, making him semi-divine. Heracles simply lifted the cruel Antaeus as high as he could, wounded him, but continued to hold him up, so he had no contact with Earth, and he died.

The French fashion and design house Chanel even created a fragrance for men called Antaeus.

The nymph Amaryllis is also connected with Earth or soil. In Greek myth, Amaryllis was a radiantly beautiful young nymph who had fallen head over heels in love with a young shepherd by the name of Alteo, who made it obvious that he was not interested in her. He decided to ask for the seemingly impossible—a new flower that had never been seen before—from any potential lover. Amaryllis consulted the oracle at Apollo's temple at Delphi. She was told to linger at Alteo's door for 30 whole nights. She was to tell him every one of those nights that she loved him, and she had to prove it by penetrating her heart with an enchanted arrow made of gold. The blood fell upon the ground where she stood for all those nights, and on the thirtieth night a red flower sprung up from the blood-drenched patch of earth when Alteo finally opened the door and his heart to Amaryllis and the new flower.

Earth is connected to the mythological origin of the Ashante people of Ghana. It is not actually known from where these people originated, but according to their legend, they emerged from a hole in the ground.

In England, there is a place called Cerne Abbas where there is a carved hill figure believed to be of Cernunnos, the Celtic hunter god. He is 180 ft in length, carries a club and features an erect phallus. He is sometimes depicted with a club and is sometimes known as 'the horned one.' He was worshipped in parts of France and in England. Cernunnos was also associated with animals and forests. The Romans believed that this giant was associated with their god Mercury, the messenger god. Because of the erect phallus, the giant has obvious associations with fertility. Some people still believe that if a childless couple spent the night on the hillside it would cure their infertility (Fig. 4.3).

There are many goddesses and even a god associated with Earth. In Egyptian mythology, Neith is the great mother goddess and the goddess of the home. Geb, unusually, a male deity, is the Egyptian god of Earth. He can be represented by a goose, as he is supposed to have laid 'the Great Egg' and has been depicted lying supine displaying his erect phallus.

There are a few goddesses that can be associated in some way with Earth from Greco-Roman mythology. Gaia is the most obvious, as the Earth goddess who came out of chaos. Uranus, the sky god, was both her son and husband. She bore the Titan dynasty of deities, including the Cyclopes, the 100-handed giants, and the monstrous giant Typhon. Gaia was the force behind many major decisions. She supplied

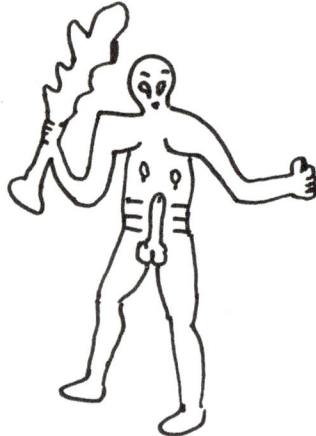

Fig. 4.3 Cernunnos was a Celtic hunter god linked to fertility. (Illustration by the author)

that infamous sickle with which Cronos sliced or shaved off the genitalia of Uranus, thus disconnecting the sky from Earth. She instructed her daughter, Rhea, to substitute a swaddling-clad stone for the baby Zeus, which saved him from being eaten by Cronos, and she presented Hera with an apple, a fertility symbol, upon her engagement to Zeus. The idea of Gaia developed into an ecological ideology that conveys the message that humans should respect the health of the planet, which matters more than any individual species presently residing on it, including the human race.

Demeter/Ceres can be associated with Earth. She is Zeus's sister and Persephone's mother. Ceres taught humankind to grow crops and to plant seeds. Vesta was the Roman goddess of the hearth and home, and Hestia was her Greek counterpart. Eurynome was the original goddess of all things. She first of all created the snake. As a dove, she laid the original egg. The celestial bodies—Sun, Moon, stars and Earth—emerged from this egg, so Eurynome was literally the mother of Earth. The concept of Earth as a mother would then include the goddesses Hera, Artemis/Diana and Aphrodite/Venus. Hera, as queen of the gods, was responsible for marriage; Artemis was the goddess of women; and Aphrodite was responsible for sustaining the human race through reproduction.

In other cultures, Papa was the Polynesian mother goddess, and in Norse mythology, Frigg, daughter and wife of Odin, was associated with love, birth and reproduction and ironically, of marriage—ironically, as Frigg also slept with Odin's brothers, Vili and Ve. The Hindu Devi contained the attributes of all the goddesses, life and death, happiness and agony, and resolution and conflict.

Cybele was a mother and fertility goddess in Phrygian mythology. She was also linked to the Moon, with its cycle being compared to her powers over life, death and regeneration. She was especially connected to Earth with a blackish stone that

Fig. 4.4 In one legend, violets grew from the blood of Attis. (Illustration by the author)

was possibly a meteorite. Her worship was believed to date as far back in history as 1200 B.C. in Asia Minor. The black stone became symbolic of the goddess and was taken to the city of Rome in 204 B.C. at the insistence of the sibyl of Cumae. Cybele was in love with a beautiful young man by the name of Attis.

It is believed that Attis was conceived perhaps even by Cybele herself from an almond. (The almond sprang from the male genitals of Cybele, who was androgynous.) Attis was an impulsive young man and cheated on Cybele with the nymph Sagaritis. Cybele discovered this and was livid with jealously and rage. She caused the unfortunate Attis to go mad, and he castrated himself. It was believed that violets grew from the blood of Attis when he severed his own genitals (Fig. 4.4). Cybele transformed Attis into a pine tree to stop him from killing himself. The pine tree, being evergreen, symbolizes immortality, and the pine cone is phallic, representing the male creative force and is sacred to Cybele (Fig. 4.5). Other variations of this myth involve Cybele marrying Attis without her father's permission that, in turn, caused her parents to murder the unlucky Attis or that Attis was castrated by a love rival. Male priests of the cult of Cybele castrated themselves! Unsurprisingly, her cult was prohibited until the end of the first century.

Demeter/Ceres was a defender of marriage despite not being married herself. Her single status gave her more freedom to take more lovers, Zeus being the most important of these despite being her brother. Demeter dared to have a brief sexual liaison with one of Zeus's children, his son by Electra. Upon encountering the lovers, the jealous Zeus flung one of his famous thunderbolts at his unfortunate offspring, which killed him immediately. She then took a human lover, Mecon, whom she changed into a poppy, a flower that became associated with her in her role as a goddess of death, although the point of turning her lover into beautiful flower was

Fig. 4.5 Pine trees were also believed to symbolize Attis. (Illustration by the author)

so that, in a way, he would never die. Their union produced Plutus, the god of abundance and riches. She also slept with Poseidon/Neptune, albeit reluctantly. She tried to escape his tempestuous clutches by transforming herself into a horse. This did not, for one single moment, stop the forceful Poseidon, who simply metamorphosed into a wild stallion. Their union produced Arion, the flying horse.

It is no accident that Bram Stoker gave the boat in which the vampire Dracula traveled from Romania to Whitby, England, the name Demeter. Dracula was accompanied by 50 cases of common earth. Soil from his home was refuge, comfort, safety and rejuvenation for Dracula. When this very earth was re-sanctified to God, not to man, then it was of no use to Dracula.

In Celtic mythology, lakes and rivers and streams were usually believed to be sacred. The Germanic Rhine Maidens were airy, female spirits who lived in water in the winter and forests in the summer. They caused the rivers to turn black, which represented their anxiety and pain when they lost their stores of Rhine gold. Ran was a Nordic sea spirit who was sometimes benevolent, while at other times malevolent, collecting sailors in her net and then abducting them.

For the ancient Greeks, each body of water was assigned a water nymph. Water nymphs were known as Naiads, and sea nymphs Nereids. In Hindu mythology, the Asparas lived in clouds and water on Earth. Mermaids are the most obvious of the water spirits; their upper bodies have faces resembling beautiful young women with long, often seductive golden hair but resembling a fish from the waist down. The Lorelei of the river Rhine, which runs from Basel in modern-day Switzerland to Bonn in Germany, are initially thought to be attractive but usually cause the death and destruction of easily seduced men.

In Greek mythology, the Sirens were flying, half-avian, half female creatures whose song was so powerfully enchanting that it banished all thoughts of home for sailors so that, in one variation, they remained on their island. In another variation, their song was so beautiful and terrible that sailors flung themselves off their ship and drowned. The famous ship, the *Argo*, carrying the Argonauts, had to pass this treacherous area of the water on their journey. Odysseus took charge of this. He got the others to secure him to the ship's mast (so he couldn't fling himself into the water while listening to the Sirens). The Argonauts were told to block their ears up with wax. As if that was not enough, Odysseus got Orpheus to sing and play his lyre so loudly that the Sirens' song could not be heard. Consequently, the *Argo* passed by safely, and the Sirens secured no victims.

Dwarfs in mythology can be seen as guarding the 'element' of Earth. They live underground in the bowels of Earth, possess precious metals such as gold and silver and jewels, which they jealously guard and, especially in Norse mythology, are master craftsmen. Four of them stand at the corners of Earth itself to hold up to the sky, rather like Atlas in Greek mythology. Dwarfs tend to live in dark places—caves and tunnels. They are afraid of sunlight (which can actually transform them into stone) and so are sometimes associated with the Underworld and the dead. In Greco-Roman mythology, Hephaestus/Vulcan, the Smith god, can be depicted as a dwarf.

According to the Bible, humans are the guardians of Earth. However, in recent times, all too often Earth and its resources are regarded as nothing more than another economic resource. People seem to be the purveyors of destruction. Humans have developed a hubristic attitude towards the planet, trying to harness its resources for our own ends, stripping rainforests of trees, genetically engineering crops, fracking for shale gas, causing the melting of the polar ice caps due to climate change, and so on.

From the 'green' or ecology movement has developed the Gaia theory, which regards Earth as alive, and promotes the idea that humans should attempt to conserve its natural resources and respect the health of the planet itself.

The planet that we all reside on and claim to know the most about still provides us with some surprises, and we are still pushing the boundaries. Our Earth has inspired much exploration, literally to the ends of Earth, a phrase going back to the days when it was believed that Earth was flat. There is something spellbinding about Earth that drives us to push the boundaries.

The 'ends of the Earth' are now regarded as the polar caps, exploration of which was begun by Captain James Cook as far back in history as 1773, when he crossed the Antarctic circle but did not make a landing. In 1908, the legendary British explorer Ernest Shackleton and his men climbed the volcano Mount Erebus (meaning "hell" or "the Underworld"), the coldest place in the Antarctic. In 1909, they planted the British flag at the furthest point south that anyone had ever been before.

Then, in December 1911, the final, unconquered place on Earth was claimed by the Norwegian explorer Roald Amundsen and his men.

The British expedition, led by Captain Robert Falcon Scott, was not so lucky. They reached the South Pole only to see the Norwegians flag already there and died on their way back. Their bodies, with letters and diaries on them, were recovered in 1912. They had tragically frozen to death in their tent. The men became national heroes in Britain and are still regarded as such, with their courage and tenacity capturing the imagination of not just the nation but the world at the time.

Captain Scott's hut remains at the Antarctic, a poignant reminder of his achievement and tragic demise. Scott and his men were prepared to embark on an epic journey, which they knew would separate them from their homes and families for three whole years. This pioneering spirit of adventure is a strong part of the human psyche with astronauts and cosmonauts alike, prepared to embark on space exploration missions that could take years, and some are even prepared to take on missions from which they will never return.

Back down on Earth, the English BBC children's' television program host Helen Skelton made an epic 5,000-mile journey across Antarctica to the South Pole by a combination of skiing, kite-skiing and cycling at the age of 28 in 2012, and in aid of the charity Sport Relief. This a continuing the legacy of Scott and Amundsen. On February 7, 2014, it was announced that the Plymouth-born Ben Saunders and his French cameraman, Tarka L'Herpinieres, trekked from Scott's famous Terra Nova hut to the South Pole hauling their own food and equipment on sleds. This is believed to have been the longest polar journey in history, at 1,795 miles on foot, taking a mere 105 days. Included among the patrons for this expedition was Falcon Scott, grandson of Captain Scott. Falcon Scott himself helped to restore his grandfather's abandoned hut on Ross Island in 2012.

In such hostile terrain, with only moonlight for months of the year and vast expanses of white, glistening ice and snow, and loneliness, it is not difficult to imagine a world of frost giants or why some people have gone mad. Antarctica has never been settled by humans, so there has never been any native knowledge, or even mythology.

It is perhaps no accident that on December 6, 2011, it was publically announced that Captain Scott's letters to his family and drawings were to go on show to the public in Cambridge. It was also announced on the same day that scientists discovered a new Earth-like planet named Kepler 22B. The newly discovered planet is 600 light years away and is more than twice the size of Earth, with a surface temperature of around 22 °C. It also orbits its parent star at approximately the same distance as Earth in relation to our Sun. Sir Patrick Moore, the late, great British amateur astronomer, suggested that it should be called Mnemo, meaning "the unknown," while English particle physicist and television host Professor Brian Cox suggested Earth 2 as its name.

Antarctica remains the coldest, most hostile and uninhabited place on Earth. We still know very little about this 5.5-million-square-mile expanse of ice. A British team intended to conduct an Antarctic exploration that included drilling through 2 miles of ice to reach a lake trapped for half a million years. Water samples would then have been taken in the hope that there would have been signs of life. If signs of life were discovered, it would have provided amazing implications for life on other planets. However, this search for life in the ancient lake, Lake Ellsworth, had to be abandoned. The crew did not get any deeper than 300 m down. Three years of planning and eight million British pounds have drawn a blank. It will be at least several years before another attempt will take place.

On September 17, 2012, it was announced to the world that Sir Ranulph Fiennes and five other men intend to cross the entire continent of Antarctica. This would be the first expedition crossing this continent on foot during the polar winter. It would be a 6-month, 2,000-mile trek across ice in temperatures reaching as low as −70 °C. The men would have to rely on the weaker reflected light of the Moon for 4 months rather than the sunlight. Sir Ranulph Fiennes was 68 years old at that time and is the world's greatest living explorer. He was the first man to trek across both polar ice caps, the first to cross the planet from pole to pole and the oldest man, at age 65, to conquer Mount Everest. He was also the first man to cross the continent of Antarctica by himself.

Fiennes has said that there is "only one really big polar record left." However, history might repeat itself, as there were rumors of a rival Norwegian expedition. Sir Ranulph's men departed for Antarctica on December 6, 2012, with the trek beginning in March 2013. This would not only push geographical boundaries but the physical limits of the human body on what has been called the coldest journey on Earth. If the human body can survive these harsh conditions it would have exciting implications for future space travel and for finding life on other moons and planets.

Sir Ranulph Fiennes is an inspirational figure. He is trying to successfully continue the British tradition of pushing boundaries and breaking records. Unfortunately, on February 25, 2013, it was announced that he had been forced to pull out of this endeavor due to severe frostbite. The expedition had to be re-planned without him. Then, on June 18, 2013, the Coldest Journey team was halted its original mission to cross Antarctica in the polar winter. It discovered that it would be perilous to try and traverse a vast crevasse field with heavy equipment, which included specially adapted Caterpillar bulldozers. Instead, it was decided the group would concentrate solely on its scientific experiments. The team did, however, spend 4 months in the dark on the Antarctic plateau before reaching its final destination of Princess Elizabeth Research Station, an amazing feat in itself. Members wore battery-powered heated clothing and specially designed breathing equipment, as inhaling air at temperatures below −60 °C (the average temperature in the South Pole) can cause permanent lung damage. Although the team did not achieve the original goal, it had acquired invaluable scientific data and survived the most inhospitable conditions on our planet.

Antarctica seems to be the most popular frontier here on Earth at the moment. The adventurer and environmental scientist Tim Jarvis has announced an important expedition recreating the 1916 Antarctic voyage undertaken by Sir Ernest Shackleton. The Jarvis expedition began early in 2013 and used similar equipment and food to that of Shackleton. In January 1915, Shackleton's ship, the *Endurance,* got trapped in the ice, eventually causing its hull to leak. The ship was used for shelter until October 1915, when they abandoned their vessel and set up camps on floes of ice. After 5 long months, they climbed aboard three lifeboats, and after 5 days of extremely cold weather in the Southern Sea, they reached Elephant Island. This island was very remote, so Shackleton and two other men traveled 800 miles to reach South Georgia Island to try to get help in the boat, the *James Caird*. They crossed the mountainous island and eventually reached a whaling station, where they got help. All the men on Elephant Island were eventually rescued and all survived.

The training ship *TS Pelican* is going to the Antarctic to support Tim Jarvis and his team. The *James Caird* boat was replicated for this expedition, which will be televised, so we will all be able to see the impossible journey that Shackleton made that required a will of iron and almost superhuman strength and stamina. Jarvis and his team reenacted Shackleton's feat complete with replicas of the clothing worn on the original expedition. The UK's Prince Harry and a team of wounded servicemen and women had intended to walk over 200 miles to the South Pole in a race against similar teams from Australia, Canada and the United States for the charity Walking for the Wounded in 2013. However, bad weather in Antarctica caused the race to be suspended. The teams later continued with the trek and completed the journey successfully with Prince Harry's team arriving at the South Pole on December 13, 2013.

A previously unnamed part of British Antarctica of 169,000 square miles was named Queen Elizabeth Land in December 2012, the year of Queen Elizabeth's Diamond Jubilee. A few days later, unsurprisingly, Argentina and Chile disputed this.

Globes, spheres showing maps of the land and sea, have been in existence since the time of the ancient Greeks, who knew much about Earth. They knew that the world was round, and Anaxagoras in the fifth century B.C. stated that he believed the Moon was like Earth in some ways. Globes themselves became the toys of the wealthy, especially in the eighteenth century. A globe of 1491 would have looked very different, without the continent of America, before the return of Christopher Columbus (1451–1506), who discovered it in 1492.

Globes still hold much fascination, especially old ones, as they are our image of the world we live in, as are maps. It is by at first imagining and then by compiling, measuring and creating maps and globes that we have formed an understanding of our surroundings.

Until the seventeenth century, it was believed that our part of the sky was geocentric, so it is no wonder that almost all mythology is connected to Earth. In the English language, our own beautiful planet is not named after a deity. Instead it shares the same name as the earth or ground beneath our feet. Earth is also one of the four elements (the others being fire, air and water) and was used by Allah and Prometheus to form humans from. Our planet was described by Neil Armstrong when seen from the Moon as "very blue and covered with white lace," probably a very comforting sight for the astronauts—home, a familiar celestial object in a deep black sky.

Earth is constantly capable of surprising us, despite our insatiable desire to know everything about our own planet and beyond.

Volcanoes

A volcano is a place or opening in Earth's crust where lava, steam, gas or molten rock from underground is either spewed out regularly or sporadically.

Volcanoes are unpredictable, sacred mountains, often regarded as symbolic of some powerful deity. The word volcano derives from Vulcan, the god of fire in Roman mythology and the name given to a fictitious planet imagined by scientists to explain the planet Mercury's strange orbit.

Volcanic eruptions on Earth have sometimes caused strange, seemingly unnatural things to occur. After the volcano Krakatoa erupted in 1883, a series of red sunsets occurred. The extensive dust from the eruption also made the Sun appear bluish and even greenish in India. The terrifyingly loud sound of this eruption was heard over 3,000 miles away. In 1927, more violent eruptions formed a new volcano. Anakrakatoa (meaning "son of Krakatoa") is growing at an alarming rate and is regularly active. The repercussions of the eruption in 1815 of Mount Tambora in Indonesia created climate changes that led to what is known as the 'year without a summer.' This understandably terrified people. Some thought this was the end of the world, and this occurrence contributed to one of the worst famines of the nineteenth century. Mount Tambora's ash blocked the Sun's rays and affected the atmosphere even thousands of miles away in Europe and Britain. This prompted writers such as Lord Byron to compose apocalyptic poetry, both beautiful and terrifying. It is believed that if the 'super' volcano Yellowstone in the United States erupted, the consequences would be even more globally catastrophic than those of Tambora, with a worst-case scenario of a little ice age.

The ancient Greeks explained the volcanic landscape of Mount Etna, Sicily, by using mythology. They believed that the gargantuan monster Typhon was buried beneath this volcano, and that his brother, the dead giant Enceladus, produced the flames with his angry breath. Through mythology, these long-dead creatures live on in people's imaginations. When Mount Vesuvius erupted in A.D. 79 in Italy, it consumed the cities of Herculaneum and Pompeii, killing around 30,000 people. At that time it might have thought that Neptune/Poseidon, who was known as the Earth shaker, was angry.

The ancient Greeks believed that Vesuvius was a place that Heracles/Hercules inhabited. The town of Herculaneum at the volcano's base was named after him. It is believed that the volcano Vesuvius formed after the ancient Somma volcano collapsed.

The islands of Hawaii are built from volcanoes, so it is only to be expected that there would be associated mythology. Hawaii consists of five volcanoes: Kilauea, Mauna Loa, Mauna Kea, Kohala and Hualulai. The volcano is personified by the destructive fire goddess Pele for Hawaiians. Only Pele can control the volcano, Kilauea's lava flow. Their mythology demonstrated that Pele was unpredictable and could erupt at any moment. She was definitely not a deity that any believer would want to even slightly annoy!

Pele became attracted to a young man called Lohiau. As he was on an island close to her, she got her sister, Hi'aka to go and collect him on her behalf.

Hi'aka asked Pele to look after her forests and her friend, Hopoe, during her absence. It took a long time for Hi'aka to discover that the unfortunate Lohiau was in fact dead. She resurrected him by using her magical powers. Tempestuous Pele was suspicious, as all of this had taken Hi'aka away for a long period of time. Without even stopping to think, Pele spewed up her fire, which completely destroyed Hi'aka's forests and killed Hopoe. When she eventually returned to Pele, Hi'aka was devastated on seeing how she had wronged her sister. Hi'aka sought solace in the arms of the attractive Lohiau. Unfortunately, Pele saw them kiss and so spewed out her molten lava, covering Lohiau which turned him to stone.

According to Hawaiian mythology, Pele lives deep within the volcano Kilauea. She is not only the fire goddess but also the goddess of volcanoes, violence and lightning. Ancient Hawaiian people offered her gifts of fruit, plants or fish to either show appreciation of her or to appease her. They believed that eruptions were gifts to the people from Pele, a way of her giving them more landmass. Pele was also supposed to have metamorphosed at times into an exceptionally lovely looking woman.

To this day Pele commands the respect of the people. Their ancestors came to the edge of Kilauea's crater to show respect for Pele. Many volcanoes, including Kilauea, have become protected sites, making it illegal to take away any lava or pumice souvenirs. It is considered extremely bad luck to take away parts of Kilauea, no matter how small, as this act could incur the wrath of the tempestuous Pele. Many 'souvenir' hunters have actually returned pieces of Kilauea, as they have experienced bad luck afterwards!

Kilauea looks like a dead landscape, black from the cooled lava, but it still has moments of activity. It has been erupting since 1983, so it cannot be disregarded. Neither can its volcanic neighbor, Mauna Loa, the biggest volcano in the world. The Hawaiian islands are an excellent example of how volcanic eruptions create land.

Iceland was formed 20 million years ago from a series of volcanic eruptions and still experiences eruptions about every 3 years. The Icelandic volcano Eyjafjallajokull spilled out volcanic ash that grounded aircraft, leaving many people stranded in April 2010. Its eruption formed two new hills named after the Norse thunder god's two sons, Magni and Mosi or Modi by the giantess Jarnsaxa.

In Germanic legend, Surt was a fire giant whose famous weapon was a blade with which he would set fire to the world at Ragnarok. After that, Surt would emerge from the land of fire, Muspell. Between 1963 and 1967, a new island that was formed from volcanic eruptions near Iceland was named Surtsey, after Surt. Some of Iceland's other volcanoes caused even more problems than those of Eyjafjallajokull. In 1783, the volcano Laki erupted. It killed thousands of residents of Iceland and even people in Britain. The sky must have turned black, and there were probably violent thunderstorms. People would have thought it was the end of the world. However, the most terrifying Icelandic volcano eruption would come from Katla, which tends to erupt every 50 years or so. The effects of an eruption can reduce sunlight, which, in turn, can alter Earth's temperature. Eruptions can produce toxic mists that kill people. However, this is the worst-case scenario, and most eruptions are not that damaging.

Mount Fuji is often said to be the world's most beautiful volcano. It is the highest volcano, out of the 109 that make up the islands of Japan. Mount Fuji is on Honshu, Japan's largest island. It has become a symbol of Japan itself and is a popular subject in Japanese art and featured in some Japanese currency. It has been regarded as a sacred place since ancient times and is believed to have been formed in 8500 B.C. It is an instantly recognizable mountain because of its beautiful symmetry. Its last eruption was in 1707.

There are different myths concerning the birth of Mount Fuji, known as Fuji-san to the Japanese people. Some maintain that it evolved from an earthquake around 300 B.C. Other imagine that an enormous carp awoke suddenly and kept on striking the sea, which in turn caused a tsunami leading to the birth of the group of islands known as Japan. However, there is another legend that maintains that Fuji was born in 1 day. A man called Visu lived in the woods with his family. One night, he woke up suddenly, startled by a thunderous sound. Visu believed that the noise was that of an earthquake. Terrified, he gathered together his family, and they left the woods as quickly as possible. He returned the following day to discover that the previously flat, dead landscape around his residence had been transformed into a mountain. He gave it the name of Never-dying Mountain, or Fuji-yama.

For years and years, many undertook pilgrimages that involved the climbing of Mount Fuji. It was believed that the Fuji goddess, Sengen, lived on Fuji's summit and that she would fling pilgrims off the mountain who were impious.

Another myth involves an adolescent boy called Yosoji. Yosoji's mother had contracted smallpox. Yosoji was afraid that his mother would die, so he called on a local magician for advice, who instructed him to find a stream at the base of Mount Fuji next to the shrine of the God of the Deep Breath. This water was apparently enchanted and would cure his mother. Yosoji did as instructed and tried to find this magic stream. However there were three different paths to the shrine, and Yosoji had no idea which one he should take.

Suddenly, out of nowhere, a beautiful young woman appeared. She led him to the enchanted stream, where he stopped briefly to drink some of the water before taking it for his mother. The woman told Yosoji to return in 3 days' time, when he would need more enchanted water. By the time he had collected water from the steam on five more occasions, his mother was completely cured, as were other villagers who had been fortunate enough to drink some of this magical water. They were all grateful to Yosoji but knew that it was really the woman they should thank. Yosoji went in person to thank her but discovered that the stream had dried up. He sat down and cried. He then turned around and saw her smiling at him. He asked what her name was. She did not answer. Then, a cloud came down and surrounded her and she floated to the summit of Mount Fuji. He then knew that the woman had been the sacred mountain's goddess.

Buddhists believe that Mount Fuji is the residence of the Buddha of All-Enlightening Wisdom. They call the mountain's summit *zenjo*.

Mount Sakurajima on one of the Japanese islands called South Kyushu is one of the most active volcanoes in the world. It started its current eruption in 1955 and is still going. Sakura means "cherry blossom," which is a hugely important plant in Japanese culture. People in Japan gather in spring to celebrate the cherry blossoms on the trees, which are beautiful, but their presence is fleeting. Sakura is a popular female name in Japan, and the cherry blossom has inspired artists and haiku poets.

The mountains of China have always been regarded as holy. People went to the mountains to seek advice from the oracles. The greatest oracle was Tai Shan, worshipped as the mountain of creation and all beginnings. The ancient Chinese people believed that its summit is a porthole or gateway for souls to come and go from. Shrines to Tai Shan exist throughout both Japan and China, and stones from Tai Shan itself are believed to be blessed.

In Masai mythology, lightning struck Mount Kilimanjaro in Tanzania. This made the mountain erupt, leading to another enchanted sky fire. After Kilimanjaro had cooled down, it left beautiful precious blue stones in the ash that we now call tanzanite.

Jupiter's moon Io is the most volcanic body in our Solar System. There is also volcanic activity on other moons, including Triton, Europa and Titan and on the other rocky planets, Venus, Mars and Mercury.

On Earth volcanic activity is usually found where tectonic plates meet or diverge. A more romantic explanation comes from the astronomer Johannes Kepler, who believed that volcanoes were the ducts for Earth's tears.

Aurora Borealis

The aurora borealis is also known as the northern lights, and the aurora australis are the southern lights. These eerie lights are the result of our atmosphere colliding with that of emissions from the Sun. The Sun sends out a continuous flow of charged particles in what is known as the solar wind. These collide with atoms of oxygen and nitrogen at the North and South poles. When solar activity increases, the likelihood of an aurora occurring increases. The aurora resemble ghostly, luminous, colored streamers moving through the winter sky. Aurorae colors include green and red from the element oxygen and pink and purple from the element nitrogen.

The Roman goddess of the dawn was Aurora, the equivalent of the Greek Eos, sister of Helios, the Sun god and, and daughter of the Titans Hyperion and Theia.

In northern skies, the winter can last for months, and from mid-October, some places such as Spitzenberg or Svalbard, islands halfway between Norway and the North Pole, there is no Sun for four whole months. In these dark, bleak, icy, bitterly cold conditions, many myths were born and passed down through generations. The aurorae still hold a strange, ghostly fascination for people and have provided the inspiration for many narratives.

In Norse mythology, the northern lights were considered to be dazzling illuminations from the shields and weapons of the Valkyries, the warrior maidens of Odin/Woden, who was the chief of the Aesir dynasty of gods. Their purpose was to ride on winged horses, cherry-picking from the dead bodies left from battles the bravest and boldest, whose souls were taken by them to Valhalla, Odin's hall in Asgard was where the gods resided.

In Germanic mythology, the Aurorae were believed to be the rays of light that the gods emitted. Gerda was an exquisitely beautiful frost giantess. The fertility god Freyr wanted to marry her, but the feeling was not exactly mutual. However, Freyr's servant, Skimir, threatened to cast a wicked spell on the charming and delightful Gerda if she refused Freyr, which would render her so hideously ugly that no man, god or giant would ever come near her again. Gerda reluctantly agreed to meet Freyr in a forest after nine nights, each night representative of a month of the northern winter.

Some Vikings believed that the Aurorae were the spirits of dead maidens. However, the Sami, who were native Norwegians, believed that they were the souls of unmarried maidens, condemned for all eternity to dance in the dark Arctic skies, trapped in the emptiness between the land of ice and snow and heaven. According to Danish mythology, the Aurorae were unmarried women or beautiful swans trapped forever in the icy depths. The Aurorae have also been thought of as a gift from the winter god Apollo.

The Inuit believed that when people die, their souls rise up to the heavens to form part of the northern lights, which consist of human souls awaiting rebirth. The lights illuminate the way for the living, and the dead communicate with the living by whistling. If the living hear the dead whistling during the aurorae, they must whistle back, and then the lights will come closer to Earth.

The Greenland Inuit believed that the souls of the dead go to the land of abundance to feast and play ball games with a skull of a walrus as the ball. To them, this ball game appears as the aurorae, and their name for them is *arsarnerit,* meaning "the ball players." Variations of myths from Greenland and the Canadian Arctic reveal that the souls of the dead produce a crackling noise when they race across the hard snowy ground of the heavens. The Point Barrow Inuit believed aurorae to be the personification of a malicious spirit, and they carried weapons to protect themselves from it. The Scandinavians regarded the northern lights as the 'light of the fox,' created by a magical fox sweeping its tail across the sky, whereas the medieval English believed that they showed God's wrath and displeasure. Even today, some Japanese believe that a child conceived beneath the northern lights will be especially intelligent.

The aurorae are not confined to Earth. They occur on the planets Jupiter and Saturn and even on Jupiter's moon Ganymede and have very recently been discovered on the planet Uranus.

The aurorae have also been regarded as the harbinger of war and of inauspicious tidings when they appear in places that rarely get to witness them. In 1939, the year that the Second World War began, the aurorae were witnessed in London. In January 2012, the aurorae were seen in northern England, Scotland and even as far south as Leeds. This could, in time, be considered somewhat inauspicious, although we now know that they are caused by increased solar activity. Coronal mass ejections (CME's), which are eruptions of bubbles from the Sun's magnetic field, interact with Earth's own magnetic field to form aurorae.

Chapter 5

Craters, Dragons, Festivals, Gods and Goddesses of Earth's Moon

"And, like a dying lady lean and pale
Who totters forth, wrapp'd in a gauzy veil,
Out of her chamber, led by the insane
And feeble wanderings of her fading brain
The moon arose up in the murky east
A white and shapeless mass.

Art thou pale for weariness
Of climbing heaven, and gazing on the earth,
Wandering companionless
Amongst the stars that have a different birth,-
And ever-changing, like a joyless eye
That finds no object worth its constancy?"

—"To the Moon" a poem by Percy Bysshe Shelley (1792–1822)

What is it about the ball of rock that sits a quarter of a million miles away that has inspired so many people over the millennia? Astronomers and scientists throughout history have fallen under its spell. Sir Patrick Moore declared the Moon as the love of his life, and the English particle physicist, Professor Brian Cox, revealed that the exploration of the Moon inspired him to become a scientist.

Sir Patrick Moore died in December 2012 aged 89 years. He was a self-taught amateur astronomer, an exceptionally talented individual with a rare generosity of spirit. He was always accessible and inspired practically everyone in the British, if not the worldwide, astronomical community. He had been fortunate enough to have met some hugely important figures in his life, including the Russian cosmonaut Yuri Gagarin and Albert Einstein. He became a member of the International Astronomical Union and was also a patron of Torquay Boys' Grammar School, Devon, England, which his *Sky at Night* co-presenter, Dr Chris Lintott, attended as a boy.

© Springer Science+Business Media New York 2015
R. Alexander, *Myths, Symbols and Legends of Solar System Bodies*, The Patrick
Moore Practical Astronomy Series, DOI 10.1007/978-1-4614-7067-0_5

Fig. 5.1 The view of Earth from the Moon. (Image courtesy of NASA)

Moore compiled the Caldwell Catalogue of 109 bright astronomical objects, which included the Eskimo Nebula, Hyades and the Sculptor Galaxy. He was a prolific writer of fiction as well as science books. He mapped the Moon, and in 1959, these maps were used by the Russians to identify parts of the dark side of the Moon from *Lunik 3*. He had been host of the television program *The Sky at Night* since 1957. He achieved a sort of immortality, not only in our memories but of a more celestial kind, as an asteroid, 2602 Moore, was named after him.

The Moon is Earth's natural satellite (Fig. 5.1). It is unusually large in relation to Earth—3,475 km (2,159 miles) across, and circles Earth every 27.3 days. We only see the Moon in reflected sunlight, as it does not produce any light of its own.

The ancient Greeks, specifically Anaxagoras in the fifth century (500–428 B.C.), knew that the Moon does not have any light of its own, its light coming from the Sun, and that it has plains and ravines similar to Earth. Ancient Greeks thought of the heavenly bodies in a physical way. The Italian astronomer Galileo Galilei (1564–1642) much, much later, in 1610, said that he thought the Moon was like Earth.

The Moon was an obvious first target for exploration, being our nearest celestial object. The race to reach it was fueled by the political rivalry between the United States and Russia. In January 1959, Russia launched a probe called *Lunik* or "little Moon." It missed the Moon and orbited the Sun. However, Russia did not give up, and *Lunik 3* was launched at a later date, which flew all the way around the Moon, showing pictures of the dark side of the satellite, the side that always faces away from us here on Earth. A later *Lunik* revealed to the world that the Moon was a cold and lifeless place, severely battered by meteorites. The late 1960s was the age of Apollo, with the Moon becoming the latest frontier for exploration and the first men walking on it. Not only was this believed by many to have been the most amazing feat of the twentieth century, but missions to the Moon, both manned and unmanned, have revealed invaluable information about our satellite.

How spectacular and mind-blowing is the mere thought of humans landing and setting foot on another world? How amazing would it have felt for Neil Armstrong, the American astronaut, taking his first footsteps on another celestial body? These footsteps could remain for thousands of years.

The various missions to the Moon, including the final manned one so far, *Apollo 17*, have revealed that the Moon is about as old as Earth itself. That rather dead-looking, heavily cratered sphere was once an orange ball of molten rock. It was first believed that lunar rocks were completely dry. More recent discoveries from probes have revealed that craters on the Moon are packed with ice from frozen water probably brought there by comets. NASA scientists in 2010 revealed that Moon dust contains water. Water has also been discovered at the Moon's south pole, but it contains quite a lot of mercury. This has exciting implications for space travel, with the possibility that rockets could be powered by this water from the Moon.

Manned Moon landings ceased in 1972 with the *Apollo 17* mission. The political and especially the economic will for space exploration is on the wane in the western world. What has happened to that pioneering human spirit? This desire to know more about our Solar System and beyond is an integral part of the human psyche. All we need is the political and economic will, difficult in what has been called recently the age of austerity. On the final day of 2011, the Chinese government announced that it has decided to make a Moon landing—exciting news for the future.

The quicker Earth spins, the more stable it is, so it is less likely to swerve or quiver. The Moon is gradually being pulled away from Earth, which will eventually cause Earth to spin at a slower rate. At the moment, the influence of our Moon and its gravitational pull act as a cosmic braking system. The Moon also affects the angle of Earth's tilt—which is 23°. Because of this angle, Earth has variable light during its year but enough to sustain life. The seasons last long enough and are gradual enough for life to adapt, which they would not do if the angle of tilt of Earth was off by even a tiny amount.

A collision with an asteroid, moon or other object in space is supposed to have slowed down the rotation of Earth and produced the tides.

The Moon could have been a body captured by Earth's gravitational pull, or perhaps the Moon and Earth were formed at the same time. However, since the Apollo lunar missions, it has become generally accepted that the Moon came about around 4.55 billion years ago when a fairly big asteroid—as big, in fact, as the planet Mars—hit Earth, and rock from both the asteroid and Earth formed a ring around Earth. This material then became consolidated and cooled down, becoming solid and producing a crust. The heavily cratered and battered surface comes from being hit by many asteroids.

The Moon is about a quarter of the size of Earth, and it is believed to possibly have a small metallic core. It has a thin atmosphere, much thinner than that of its parent planet. Because of its thin atmosphere and no wind to cause erosion, its textured and cratered surface has remained almost the same for two billion years; that's also the reason that Neil Armstrong's footprints might remain for millennia. This surface contrasts with that of Earth, where it is difficult to tell where the impact craters are, as many have been covered over.

The Moon's seas (which are not watery in spite of their name), or maria, are smooth, as the lava that created them was very runny. The surface of the Moon is pockmarked with craters from meteorite or asteroid impacts. Some of the smaller craters, usually of around 1 km in diameter, have been given female names: Edith, Mary, Susan, Grace, Ruth, Rosa, Isabel, Stella, Shahinaz, Mavis, Kathleen and Melissa. Other smaller craters have been given male names: Ian, Michael, Jose, Yoshi, Ravi, Samir, Kasper, Ivan, Taizo and Osama. Michael could also be a reference to the archangel Michael, whose abode is supposed to be on the Moon. Other, larger craters can be identified as names of astronauts, cosmonauts and those connected to space exploration. These include Aldrin, Armstrong and Collins craters, which were named after Buzz Aldrin and Neil Armstrong, the first humans to walk on our satellite, and Michael Collins, who remained in *Columbia*, the U. S. command module, when Aldrin and Armstrong flew on to the Moon in the Eagle lunar landing module on July 21, 1969.

Yuri Gagarin was a cosmonaut and the first man to fly in space in 1961. William Anders, Frank Borman and James Lovell were the first men to orbit the Moon, on *Apollo 8*. Other astronauts to circle the Moon—Alan Bean, Eugene Cernan, James Irwin, Edgar Mitchell, David Scott, John Young, Alan Shephard and Harrison Schmitt—do not yet have lunar craters named after them. The Russian spacecraft designer Sergi Korolov (who designed *Sputnik 1*) has a lunar crater named after him, as do Dmitri Dmitrevich Maksutov, the Russian astronomical optics inventor, and Eugene Merle Shoemaker, who taught Apollo astronauts field geology. The Apollo crater is named after the Apollo space program but is also the name of the ancient Greek god of the Sun and light.

Other craters are named after characters from mythology. Chang' Ngo is the Chinese Moon goddess; Mercurius is named after the messenger and trickster god, Mercury; Daedalus is the inventor who made fake wings for himself and his son, Icarus. Icarus, despite warnings, flew too close to the Sun, with the heat melting the wax that held together his wings and causing him to drown. The Artemis Crater is named after Artemis/Diana, twin sister of Apollo and the goddess of women. The Dionysus Crater is named after the god of growth and change who gave wine to mortals. The Atlas Crater is named after the Titan god who was punished by Zeus by having to carry the heavens on his shoulders. Hercules was a demi-god, son of Zeus. He was famous for his amazing physical strength. He captured the hellhound Cerberus, was part of the Argonauts expedition to find the Golden Fleece and released Prometheus from his terrible punishment. Endymion was the Moon goddess and Selene's lover, an appropriate name for a lunar crater. Menelaus was one of Helen of Troy's five husbands, from whom Paris stole her. He was an Archaean chieftain and brother to Agamemnon, who led the Greeks in the Trojan War. The crater Osiris is named after the Egyptian god of the underworld and of plant life, and the Isis Crater is named after the Egyptian mother goddess who was both wife and sister to Osiris.

Some Moon craters are named after famous explorers, people who pushed the boundaries traveling to unknown places, some under extreme conditions. The British hero Ernest Shackleton (1874–1922) and his men traveled to the furthest point south on the Antarctic in 1909. The Norwegian explorer Roald Amundsen (1872–1928) was the first man to reach the South Pole, although the British explorer Robert Falcon Scott (1868–1912) also reached the South Pole but perished on the way back. Marco Polo (1254–1324) was a Venetian merchant who traveled to China and India and even reputedly met the Mongol Chinese emperor of the Yuan dynasty, Kublai Khan. He returned to Venice years later with Chinese silk, gold and what seemed like wildly imaginative and incredible stories. He has a lunar crater named after him, as does Christopher Columbus (1451–1506), whom he apparently inspired. Columbus was a famous Genoese explorer who is credited with the discovery of the Americas. His lunar crater is known as the Colombo Crater.

The Montgolfier brothers pushed the boundaries skywards. They invented the Montgolfier hot air balloon, and in 1783 two passengers boarded one such balloon on the first manned balloon flight, which paved the way for planes and manned space travel. They made our desire to fly possible. The Alexander Crater is named after Alexander of Macedon (356–323 B.C.), believed by some to be the son of a god rather than Philip of Macedon. Alexander was taught by the famous Greek philosopher Aristotle. He defeated the Greeks and overthrew the Persian Empire. The appropriately named Peary Crater is situated on the far side of the Moon and is the large crater that is closest to the north pole. Robert Edwin Peary (1856–1920) was an American polar explorer, the first man to reach Earth's North Pole on April 6, 1909.

There are many, many lunar craters named after famous mathematicians, astronomers, scientists, philosophers and people whose discoveries were revolutionary. Plato (429–347 B.C.) was an ancient Greek philosopher, scientist and mathematician, as was Aristotle, who devised an Earth-centered universe, updating Eudoxu's earlier version. Aristarchus is a bright, young crater named after an ancient Greek astronomer who tried to calculate the size of the Moon back in the third century B.C. Anaximander (610–546 B.C.) made mechanical observations about the universe as he saw it. He was the first man to predict an eclipse and claimed that nature was ruled by laws. He looked up at the sky and compared the heavens to what happens on Earth. These incredible discoveries have gained him a place in history and a Moon crater named in his honor.

Anaximenes (585–528 B.C.), another ancient Greek philosopher, produced a more naturalistic observation, claiming that the rainbow was produced when the Sun's rays fell onto dense, thick air. Anaxagoras (500–428 B.C.) conceived a physical account of the heavenly bodies. He believed that the Sun was a molten mass or a red-hot stone. He said that the Moon had plains and ravines on it, like Earth, and that its light was not its own but came from the Sun. These ideas and observations were incredible at the time and have made him an important figure in the history of cosmology, earning him the accolade of having a lunar crater named after him.

Hypatia (A.D. 350–415) taught and worked at the library, the center of learning, at Alexandria in Egypt. She became famous as a philosopher and mathematician of her time and rightly deserves a lunar crater in her name. Eratosthenes (276–194 B.C.) deserved his lunar crater, as he made the first maps of the world and used mathematics, more specifically trigonometry, to calculate the size of our planet. Archimedes (287–212 B.C.) is probably most famous for shouting 'Eureka' when jumping out of his bath after discovering what is known as the Archimedes principle, or law of hydrostatics. Pythagoras (580–500 B.C.) is probably most famous for his theorem about triangles. He also thought that all the celestial bodies were spheres. The Euclides Crater is named after the ancient Greek Euclid (c. 325–265 B.C.), famous for giving us Euclidean geometry and perspective and spherical geometry, the latter of which he applied to astronomy. Finally, the Ptolemaeus Crater is named after Clauduis Ptolomaeus, or Ptolemy (A.D. 90–168), who was interested in the motions of the planets and stated that they moved in circles within circles.

The Euler Crater is named after the Swiss-born Leonard Paul Euler (1707–1783), who moved to St. Petersburg in Russia, where he was known as the 'Mozart of Maths,' whereas the Gauss Crater was named after Karl Friedrich Gaus (1777–1855), known as the 'Prince of Mathematics.' He invented imaginary numbers, regarded as the key to quantum physics. The Pauli Crater is named after Wolfgang Pauli (1900–1958), who came up with the Pauli exclusion principle, which stated that every electron in the universe changes its energy levels constantly and everything is connected to everything else. He also predicted the existence of the neutrino. Johannes Kepler (1571–1630), the German astronomer and mathematician, got a lunar crater in his name for conjecturing that the gravitational pull from the Moon caused tides here on Earth. Albert Einstein (1879–1955) was an obvious candidate for a lunar crater name, with the profound effect that he had on science with his Theory of Special Relativity. Einstein also proved that light can travel through empty space and that nothing can travel faster than the speed of light. The English astronomer and host of the television program *The Sky at Night*, Sir Patrick Moore, discovered this particular crater.

The Mendeleev Crater is a tribute to Dimitri Mendeleev (1834–1907), the Russian scientist who created the Periodic Table of Elements. John Flamsteed (1832–1907) was the first Astronomer Royal in England. He was under orders to make sky observations that were relevant to navigation, as Britain was a seafaring nation. Norman Lockyer (1836–1920) was an English astronomer and scientist believed to have discovered the gas helium. In addition to having Moon craters named after them, John Flamsteed has an astronomical society named after him based in Greenwich, England, and Norman Lockyer has an observatory named after him in Sidmouth, Devon.

William Herschel (1738–1822) became the king's astronomer. He discovered Uranus and built new telescopes that used metal mirrors rather than glass lenses, capturing much more light from the stars. Sir Isaac Newton (1642–1727) discovered gravity and invented calculus as way of describing things, such as planetary orbits, using mathematical language. The Descartes Crater is named after the French philosopher and mathematician Rene Descartes (1596–1650), perhaps most famous for his philosophical statement *Cotigo ergo sum,* or "I think therefore I am." However, he also combined algebra and geometry, leading to equations for, among other things, ellipses. (Mercury has a weird elliptical orbit). The Maxwell Crater is named after the Scottish James Clerk Maxwell (1831–1879), who revealed that light is an electromagnetic wave. The Planck Crater is named after Max Karl Ernest Ludwig Planck (1858–1947), who is famous for Planck's constant, which the quantum of action (minimum amount of any physical entity in an interaction) in quantum mechanics.

The Becquerel Crater is named after Henri Becquerel, who discovered radioactivity. The Ohm Crater is named after Georg Ohm (1789–1854), a German physicist who is famous for Ohm's law of electricity, $V = I \times R$. Ohms are also a unit of measurement for electrical resistance. Enjar Hertzsprung has a crater named after him. He was a Danish chemist and astronomer (1873–1967) who created the Hertzsprung-Russell diagram to plot stars and their color and temperature. The Kuiper Moon crater is named after the Dutch-born U. S. astronomer Gerard Peter Kuiper (1905–1973). The Kuiper Belt is also named after him, and he discovered Uranus's moon Miranda and Neptune's moon Nereid. The Viviani Crater is named after Vincenzo Viviani, a pupil of Galileo. The Hubble Crater is obviously named after Edwin Hubble (1889–1953), the American astronomer who discovered other galaxies beyond our Milky Way, among other things; NASA's Hubble Space Telescope is also named after him.

There are a few other lunar craters that are named after writers such as the science fiction writers Jules Verne and H. G. Wells, as well as the Chaucer Crater named after the English writer and poet Geoffrey Chaucer (1343–1400), famous for writing *The Canterbury Tales.* The Dante Crater is named after Danti Alighiieri (1265–1312), an Italian poet famous for his *Divine Comedy.*

There are many other named lunar craters not mentioned in this book.

There is something enchanting and magical about our beautiful, lonely Moon that bathes our Earth in a romantic, silvery glow of light reflected from the Sun. The Moon has captured the imagination of so many cultures and peoples. The sight of the Moon's beautiful face on a clear night never fails to fill people with awe, wonder and inspiration. As our nearest heavenly body and visible with the naked eye, there is much mythology concerning the Moon.

The Moon, an almost constant celestial presence and frequently visible in the daytime, has had an enormous influence on China, its people, culture and customs and, more recently, on space exploration. In 2007 China launched its first lunar orbiter, *Chang' E1,* named after its Moon goddess. Unsurprisingly, a tiny crater on the Moon itself is named *Chang Ngo.*

It seems inevitable that the Chinese will make a manned Moon landing in the near future. On December 14, 2013, China successfully landed an unmanned spacecraft on the Moon's ancient surface. The robotic rover, which left its landing module the following day, is called *Yutu,* or "Jade Rabbit." The name was chosen from an online poll of 3.4 million voters and involves a character from Chinese mythology.

The Moon holds some valuable resources, such as silver and a chemical called helium 3, found in lunar soil and which is an efficient nuclear fuel. It is estimated that our satellite contains more than a million tons of helium 3. This isotope would be used in nuclear fusion reactions. Nuclear fusion uses nuclei to generate energy. The advantage of helium 3 over other isotopes is that it does not give off pollution or radioactive waste. It is estimated that 25 tonnes of helium 3 could provide enough energy to power the United States for an entire year! This Moon landing is both scientifically and symbolically important, confirming China's status as a global superpower as well as involving ancient mythology with the choice of the rover's name. China also plans to send a manned mission to the Moon by 2025, significantly before the United States returns.

In ancient China, the emperor was believed to be divine, the son of heaven itself! He almost had to prove this by being able to predict sky events, which were believed to mirror Earthly happenings. Therefore, emperors employed astronomers who observed sky events and recorded them in great detail.

The Chinese astronomers divided up the sky in a seemingly complex but methodical way. First, it was broken up into five palaces, or *gong-o,* representing a middle region and the four main points of the compass. The central region was believed to include a central image of the emperor himself, cementing his divine status, in addition to some family members. Constellations here include, for example, 'The Prince.' In addition to the constellations in the four cardinal directions, 28 lunar mansions or lodges were designated in the sky. Of time using early evening sky objects such as constellations, lunar lodges or mansions. The four directions each include 7 of these 28 mansions. The directions include:

- The north, and winter, represented by the black tortoise, an obvious symbol of long life (Fig. 5.2). Associated with water.
- The east and spring, represented by the blue dragon, symbolic of good luck and kindness. It is connected with wood.
- The west and the autumn, represented by the white tiger, symbolic of protecting and watching over both the living and the dead. It is connected with metal.

In ancient China one emperor was credited with being the father of Chinese astronomy, with the lunar calendar being developed during his reign with his brothers, who are all represented by one of the 12 astrological signs. This lunar calendar was designed to be very practical here on Earth by providing a framework in which farmers could work out when to sow and gather their crops.

Fig. 5.2 In China black tortoises represented north. (Illustration by the author)

The yin-yang principle is central to ancient Chinese life, philosophy and even the diet. It was all about balance. Yin is female, cooler, passive and lunar, the opposite of yang, being male, more active, hotter and associated with the Sun.

The Moon has influenced Chinese culture in ways that might seem less obvious. The traditional Chinese arch-shaped bridge is designed to resemble the Moon when it is reflected in water. The two arches, one solid and one reflected, together make a complete circle, also an example of the yin-yang principle. The most beautiful women were selected for their round, flattish faces, painted white to resemble the Moon, with eyebrows plucked into perfect, delicate arches, like crescent Moons. It was one such imperial courtesan, Yao Niang, who is to blame for the barbaric and now banned practice of foot binding, which involved the breaking of girls' feet and unnaturally creating small crescent feet by tightly binding them with bandages, to make them look smaller and more like the Moon! This took place from around the year 1000 until 1911. The Moon with its hare is also 1 of the 12 ornaments that appears on Chinese imperial robes. The robes have buttons on them; the larger ones supposedly symbolize the Sun and the Moon.

Each year, some Chinese people celebrate the Moon or Lantern Festival *(Zhong qui Jie)*. It occurs in mid-autumn on the fifteenth day of the eighth month in the Chinese calendar (September or early October). It is one of the most important of the Chinese holidays, dating back from over 3,000 years ago to the Moon worship in the Shang dynasty. Farmers celebrate the end of autumn, and families gather together, originally to celebrate the beautiful, bright harvest or full Moon. The Moon was believed to increase the harvest, with the Sun believed to be a threat to it in China, hence the importance of the Moon goddess over solar deities. To celebrate, fireworks are set off and people eat special Moon-cakes, some of which are filled with seed paste. Commercial Moon-cakes often show representations of the Moon goddess, Chang'e, floating or drifting up to the Moon; during the festivities, tiny hare images can be given as sacrificial offerings, and porcelain figures of white rabbits are made specially for this occasion (Fig. 5.3). Others choose to celebrate by carrying lanterns, taking part in fire dragon dances or burning incense.

Fig. 5.3 The Lantern Festival in honor of the Moon is celebrated in part through offerings in the shapes of hares. (Illustration by the author)

Some Chinese people believe that the Moon festival commemorates the Chinese victory in overthrowing the Mongol rulers in the fourteenth century. The Chinese rebel leader, Zhu Yuanzhang's advisor, devised an ingenious plan to attack and defeat the Mongol on the same day as the Moon festival. This wily advisor had recognized that the Mongols chose not to eat Moon-cakes. He sought permission from the Mongol rulers to give out the celebratory cakes to the Chinese citizens. The clever plan involved hiding a clandestine message, commanding the Chinese to all attack the Mongols on the day of the Moon festival. The Mongol government was thus attacked and overthrown, and the Chinese Ming dynasty was founded.

Many stories are told at the Moon festival. One of the most popular is "The Cruel Emperor and the Moon." In this particular tale, there lived a cruel emperor, hated by his own people. Like all megalomaniacs, he wanted to live forever, so he forced his servants to search high and low for the infamous tablet or pill of immortality. When it was eventually discovered, the unpopular emperor announced to his people that he intended to swallow this enchanted pill. The people were truly dismayed on hearing this devastating news. In order to stop this horrific event from happening, the emperor's young wife stole this magical pill and took it herself. She then found herself floating up to the Moon. This unselfish act made the people very happy and relieved. Even today, when the full Moon shines, it is believed that the emperor's wife, Chang'e, can be seen in the Moon, smiling down from her perch in the sky.

In another version, Chang'e was not so altruistic and desired immortality as much as her husband and so swallowed it all herself, resulting in her rising upwards to the Moon. She achieved her desire to immortality, but it came at a price, as she lost her beauty and was transformed into the cosmic toad, with only the company of a hare pounding a mortar and an old man chopping at a cassia tree. In yet another variation of this myth the Jade Rabbit and Chang'e crush medicine together for the gods, and in a similar story, Chang'e herself is, in fact, the Jade Rabbit. Chang'e is

Fig. 5.4 Chang'O as a moon goddess transformed into a Moon-toad. (Illustration by the author)

also known as Chang'O, the Chinese Moon goddess who was changed into the Moon-toad as punishment for the theft of the elixir, or pill of immortality (Fig. 5.4).

In China, jade was the most valuable and important of mineral stones, with its opaque but light-reflecting qualities. Objects made for religious or ceremonial rites and jewelry were crafted in precious jade as far back as Neolithic times. Carved creatures such as fish and dragons were created from jade later, during the Bronze Age. The enduring and seeming permanence of these creatures and ornaments contributed to the belief that jade contains some kind of elixir of immortality. This links to the Jade Rabbit in the Moon story. Jade charms against evil were buried with the dead.

Some tombs of minor kings and princes containing jade have been excavated in China from around 204–111 B.C. The equivalent to the ancient Egyptian mummy case in China was a jade suit. This consisted of many thin slices of jade constructed into a suit completely covering the entire body, including the face, originally stitched with golden thread. This would have looked both ghostly and terrifying yet spectacular, showing off the true beauty of the thin slices of light-reflecting jade. The body would have been put into this jade suit and then into a coffin in a tomb.

Jade was used to make the body suits because it was believed to preserve the body (and contain that elixir of immortality). It was also supposed to repel demons. Like the Egyptians, tombs often included gold, jewels and perhaps even more significantly, pestles, mortars and minerals, possibly including jade that could be used ground up as a powder or dissolved into a solution and used to cure illness or even to extend life itself, once again linking this belief to the story or myth of the Jade Rabbit. Even today, jade is believed to have healing and harmonizing properties and symbolizes purity. It is a protective stone and is believed to draw auspiciousness to itself.

The markings on the full Moon, at a particular angle, resemble a rabbit or a hare. Some have suggested that the Moon's darker areas, or maria, resemble leaping

Fig. 5.5 Unlike the 'man in the Moon' of the West, Chinese culture referred to a 'hare in the Moon', and hares have been associated with the Moon there ever since. (Illustrations by the author)

hares. Rabbits or hares are a popular lunar symbol, as they sleep during the day and are wide awake during the night. In China, the hare in the Moon, which according to some stories, pounds cinnamon sticks in its mortar with its pestle, is symbolic of a long life. The hare or rabbit are yin, lunar creatures (Fig. 5.5). A person with a hare's head with six boys around is a pictorial symbol of a wish that is made at the lunar festival for children, in the hope that they will have a smooth progression in the civil service.

The Moon goddess, Chang Xi, was the caring mother to 12 Moon daughters. She sent them out, one by one, up into the sky so that they could shine down on Earth at night. Chang Xi dutifully waited for every 1 of her 12 daughters to return in the morning. She was so caring and considerate that she went without sleep, just like any mother, until all her daughters had returned to her. When each one returned, she took care of her, bathing her tenderly in the Eastern Sea and then carefully hanging her daughter to dry on the mulberry tree.

Other stories are traditionally told at the Moon Festival. One such tale is "The Jade Rabbit." In this fairy legend, the Moon itself becomes a rabbit or hare. Here, three wise fairies metamorphosed into sad, poor old beggars. They each asked for food from three different animals—the monkey, the fox and the rabbit. Both the monkey and the fox willingly donated food to the hungry old men. However, the rabbit did not have any food to offer them. He unselfishly offered himself as food. He leapt into the roaring fire, so that he would be properly cooked. The wise fairies were so impressed with the willing self-sacrifice of the rabbit that they decided to allow him the great honor of living in the Moon Palace, where he became the Jade Rabbit, jade being associated with immortality. This rabbit can be seen in the radiant disc of the full Moon, so the Moon does become a rabbit. This story is very similar to one in Indian mythology, where the hare offered himself up as food for a Brahman who had nothing to eat. In this case, the kind offer was refused, but the outcome was the same, with the unselfish hare being elevated up to the Moon in recompense.

There is yet another Moon Festival story that is a version of the Chang'e tale. This is an incredibly ancient myth, originally pre-dating the Bronze Age, so it is therefore unsurprising that there are different variations. Once, long ago, Earth had not one but ten different suns to shine their radiant light down upon Earth at different times. However, one day, this system all went very wrong, with all ten of the suns shining down upon Earth at the same time. This heat was far too intense and damaged some parts of Earth. However, our planet was saved from scorching to death by an unlikely ally, the famous archer and harsh ruler, Hou Yi. His solution was to shoot down nine of the ten suns in the sky, restoring balance to the cosmos. This same man also managed to steal the very substance of life or immortality from one of the goddesses, which he had every intention of consuming. He did not actually manage this because his beautiful wife, Chang'e, got there first. She consumed this magical substance to save the people for a potentially immortal and harsh ruler. She then ascended up to the Moon. As Hou Yi had genuinely loved his wife, he did not even attempt to shoot the Moon down from the sky.

Another Moon Festival story is of the man in the Moon, Wu Kang. Wu Kang was a man who could never seem to stick to anything for long; he tried his hand at many professions but seemed unable to concentrate for very long. One day, he awoke and decided to set his sights very high. He would try to become immortal. He set off towards the mountains where he lived and even managed to persuade an existing Immortal to show him how to use herbs for medicinal purposes. Wu Kang soon became bored and wanted to know how to do something new. He was then shown how to play the game of Chess but soon grew tired of it. Next, he was handed the precious scrolls of immortality so that he could learn from them, a somewhat lengthy task. Predictably, Wu Kang lost both his concentration and enthusiasm very quickly, so he asked to journey to somewhere new. The Immortal had had just about enough of the annoying Wu Kang, so he directed him to the palace on the Moon and gave him a task to perform. He had to chop down the enormous cassia tree before he would be allowed to come back down to Earth. Wu Kang worked very hard, but the tree grew back every time it was chopped down, so Wu Kang remains up there in the Moon. A case of, be careful what you wish for, as you might just get it, Wu Kang achieved his goal of immortality, but it came at a price.

Another ancient Chinese Moon myth involves Yue Lao, also known as the Old Man in the Moon, who teamed up pairs of newborn baby girls and boys and tied their legs with an invisible thread. The pair were supposed to then feel an inexplicable, unavoidable attraction to each other when they became adults, which inevitably resulted in marriage.

There are other Man in the Moon stories from other cultures. Some Europeans believe that a man was nailed to the Moon as a punishment for particularly terrible crimes. Others believe that the Man in the Moon was Cain with a dog and a thorn bush, the thorn bush being symbolic of the fall of man and the dog believed to be the 'foul fiend.' A dog will be unfazed by corpses or dubious morality, thus making it the perfect companion to Cain.

Cain holds the dubious honor of being the first murderer in the Bible. Cain, the first born son of Adam and Eve, became a farmer, while Abel, his younger brother, became a shepherd. They both presented offerings to God, with Cain giving some crops and Abel some newly born lambs. God preferred Abel's gifts. This made Cain feel angry and jealous of his brother. When his anger reached the boiling point, Cain murdered Abel. God, being omniscient, knew this and was not exactly pleased. He then cursed Cain, ensuring that the ground would no longer give forth crops to him and that Cain would be doomed to wander Earth for the rest of his life. God also placed the mark of Cain upon him, which ensured that nobody would kill him to put him out of his misery.

Some believe the Man in the Moon is Endymion, taken to the Moon by the goddess Diana. The man, to some, is leaning on a fork and carries a bundle of sticks picked up on a Sunday, the Sabbath, supposedly the day of rest.

A similar old German myth tells of an old man who stole the cabbages that belonged to his neighbors. He was discovered and was banished to the Moon, where he has to carry his cabbages forever. There are two other variations of this story. In one, the man steals not cabbages but sheep, and in the second, also an old German story, the man is cutting sticks on a Sunday, so he is punished by being taken to the Moon after being given one of two terrible choices, of being frozen on the Moon or of burning on the Sun. He chooses the former. To the Oto Native Americans, the Man in the Moon was a hero, a brave young man by the name of Running Antelope. He was married to a beautiful young woman, whom the chief of the village wanted for himself. The chief decides to kill Running Antelope so he can have his wife and runs after him until they get to a lake. There, Running Antelope asks the water spirits to protect him from the murderous chief. They produce a huge jet of water that sends him directly up to the Moon.

The Moon is associated with fertility and therefore with rabbits and hares, who are famous for breeding! The gestation period of a hare also corresponds with the lunar cycle. There are tales of rabbits, hares and the Moon in other cultures. In Russian legend, the quail appears, representing the Sun, with the hare as lunar, and in Aztec symbolism, the Moon is a rabbit or a hare.

In Aztec mythology, the Moon shone as brightly as the Sun until one of their deities flung a rabbit at the face of the Moon, where it remains. In Mexico, there is a story that is very similar to one in both Chinese and Indian mythology. Their god, Quetzalcoatl, started off as a mere man. He traveled for a long period of time and eventually became ravenously hungry. He encountered a rabbit at this point who, most generously, offered to allow Quetzalcoatl to eat him. For the Aztecs, it was therefore Quetzalcoatl who caused the rabbit to rise up into the Moon, giving the creature immortality.

In Cree Native American mythology, there is a story involving Rabbit's desire to ride the Moon. Rabbit persisted in his desire, much to the annoyance of the other creatures. Eventually, Rabbit managed to persuade the Crane's legs to help him, and Crane felt himself getting heavier and heavier, which is how the Crane's legs became stretched and elongated. When Crane had flown to the Moon with Rabbit, the furry, long-eared creature reached up to Crane's head with his paw, which was bleeding. This is why cranes have a red patch on their heads. Crane flew back down to Earth, but Rabbit enjoyed riding the Moon so much that he is still there.

In South African Bushman, or San, mythology, the hare annoyed the Moon. This is a complex story. It began with the rainbow, Kwammang-a, losing part of his shoe. The creator god, Cagn, picked it up and put it into water. This eventually morphed into the first eland (also known as a taurotragus or antelope). Cagn enjoyed looking after the eland. He taught it to come to him, gave it the name Kwammang-a's shoe-piece, rubbed it with honeycomb, making it look beautiful and shiny, and gave it honey to eat, making it big and strong. Cagn was so proud of the beautiful eland that just looking at its beauty made him sing with elation.

Kwammang-a became curious about where all the honey was going. He got his son, Ni-opwa, to follow Cagn, and so discovered the beautiful eland. Kwammang-a murdered the beautiful creature and cut it up. Cagn saw his lovely eland being cut up. He was very upset and cried over this. Cagn, the creator god, then got the eland's gall bladder and made a hole in it by poking it with a stick. Darkness came out and completely engulfed the world. Later Cagn decided that the world needed some light, so he threw one of his shoes up into the sky, becoming the Moon. Therefore, the Moon walks across the sky like the shoe that he was created from. When the Moon becomes full, the Sun stabs him in the back with a knife, so that he gradually deteriorates, until only his backbone remains. Then, the Moon is gradually reborn again. The Moon had wanted the people of the San to be reborn after death, just like him, until he was angered and irritated by the hare.

An animal person in the form of a hare was devastated by the death of his mother. He kept on saying that his mother wouldn't ever return. The Moon kept trying to comfort him by telling him that she would return. This went on for a while, the hare refusing to believe the Moon. The Moon then become angry and slapped the hare and cursed him. Then, still furious, the Moon announced that men would ever after die and would not return. Thus, death began.

The Man in the Moon is sometimes believed to be Endymion. In Greek mythology, Endymion was a beautiful young shepherd. Selene, the Moon goddess, took him as her lover, and together they had 50 daughters. In another version of this myth, Zeus is believed to have given this exceptionally handsome young man the privileges of immortality and of eternal youth, because he was permitted to sleep on Mount Latmus, which is where the equally beautiful Selene floated down from the Moon to have sex with him every night.

All classical Greek goddesses can be considered lunar, as the Moon is associated with women, the feminine principle and also with virginity. One of the twelve Olympian deities was Artemis (Roman: Diana), the twin sister of Apollo, the god of light and the Sun. Artemis was the virgin goddess of women and birth. She was also a goddess of the wild and animals, enjoying the freedom of wandering the forests with her nymphs.

The unfortunate Actaeon managed to upset Artemis. He was out hunting one day. He was a gifted hunter and had been taught by none other than Chiron, the wise centaur. While out with his men and dogs, he accidentally discovered an incredibly beautiful woman, bathing naked in a river. This woman was the goddess Artemis, and she caught Actaeon looking at her. In another myth, Actaeon boasted that he was a more skillful hunter than Artemis herself. Whichever myth you prefer or believe, the outcome is the same—Artemis grew angry with Actaeon. In her fury, she transformed the unlucky Actaeon into a handsome stag. He was then hunted down, attacked and eaten by his own dogs. The faithful dogs were devastated at the horrible death of their master. Chiron consoled them by creating a statue of Actaeon that looked just like their dead master, as if part of Actaeon himself lived on through this realistic painted carving. The bow itself, an attribute of Artemis/Diana, is a symbol of the crescent Moon.

Another Greek myth involving the virginal goddess Artemis concerns Orion, for whom a constellation is named. Orion was famous for being a skilled hunter. He also just happened to be devastatingly gorgeous, the son of both Euryale and the god of the sea, the Earth shaker, Poseidon. Originally, Orion had intended to avenge a wrong, but when he met the lovely Artemis, who loved to hunt and chase animals as much as he did, he forgot about this quest. Instead, Orion and Artemis went off to hunt. However, the goddess of the dawn, Eos, had fallen in love with him, and they had slept together, the cause of the rosy pink blush that was to appear at dawn. Apollo, protective twin brother of Artemis, knew about this and did not want his sister to get hurt or used by the incredibly handsome young man, Orion, so he interfered by persuading the Earth goddess to make a giant scorpion to chase Orion.

Orion bravely fought the huge scorpion. First he tried to shoot it with his arrows and then he tried to wound the thing with his sword, all to no avail. He tried to escape by jumping into the sea and swimming towards Eos's domain, desperate for her to keep him safe. Then, Apollo interfered again by telling his sister that Orion was a villain and challenged her to shoot him. She was an excellent shot and hit her target with an arrow through the head. Orion was obviously killed. Artemis soon regretted her actions and begged Asclepius, Apollo's son and the god of medicine, to restore him to health, which he started to do, but he was hit by a deadly thunder-bolt from Zeus. Orion remained dead. Artemis then sent the handsome body of Orion up into the stars, becoming the constellation Orion, but the scorpion was also raised up to the firmament, where it forever chases Orion the Hunter.

Agamemnon, the leader of the Achaean army (who fought against King Priam of Troy in the Trojan War), was a foolish man. He was with his army, which had gathered at Aulis and were preparing for an assault on the city of Troy. While there, he shot a magnificent stag and boasted that he was a better shot than Artemis herself! Not a good idea, bearing in mind Actaeon's gruesome death and the fact that he was at war. Artemis was not pleased. She took out her revenge on the Achaean fleet by stopping the winds from blowing. This meant that the ships were stuck, and they were unable to sail anywhere. Not content with that, Artemis made Agamemnon sacrifice his daughter, Iphigeneia, to her. Artemis, however, felt sorry for the innocent girl and saved her from sacrifice by covering her in clouds. She was

Fig. 5.6 The owl, as a nocturnal bird, is associated with the Moon. It also represents Athena or Minerva. (Illustration by the author)

then made chief priestess at Artemis's temple in the Tauric Chersonese, the only human permitted to touch the sacred ancient wooden representation of Artemis herself. Although Iphigeneia was saved, Agamemnon was unaware of this, as Artemis had placed a body of a deer in the place of the girl's corpse.

Athena/Minerva, goddess of war and wisdom, another virgin goddess, is also connected to the Moon. The owl, symbolic of wisdom and learning, is one of her attributes. Being a nocturnal creature, the owl is also lunar. Athena was Zeus's favorite child, born as an adult even wearing armor, from the head of Zeus. (He had swallowed the pregnant Metis, a Titan goddess of wisdom, Athena's mother.) Athena took the side of Achilles, Agamemnon and Menelaus in the Trojan War along with Hera and Poseidon. This side was eventually victorious. Athena also provided assistance with the construction of the famous ship, the *Argo,* in which Jason and the Argonauts traveled on a potentially perilous journey to Colchis to steal the magical Golden Fleece (Fig. 5.6).

Athena was in direct competition with Poseidon/Neptune for patronage of the new city of Attica. Both deities offered the people of the city bribes or incentives. Poseidon gave them a stream that ran into the sea and Athena gave them an olive tree. The people decided that Athena's incentive was the more useful and would provide olives, wood and olive oil, and so they chose her to be their patron. She is also, more obviously, the patron of the city of Athens, which bears her name.

Athena is also associated with the Moon because she is a goddess of crafts, especially of spinning and weaving. This is shown in the story of Arachne.

There was once a beautiful young woman by the name of Arachne who was an amazingly talented spinner, weaver and embroiderer. She created pictures that were so spectacular that the nymphs frequently left their natural habitats to watch her work. Unfortunately, although very talented, she was boastful, and bragged that there was nobody as talented as her at embroidering. People thought that the goddess Athena/Minerva must have taught Arachne and told Arachne this. Instead of taking this as a compliment, she even declared that she was better at weaving than Athena/Minerva herself! A somewhat reckless thing to say. This hubristic declaration did not escape the goddess's ears. Athena/Minerva disguised herself as an old crone with the intention of merely advising the impulsive young woman. As she watched Arachne working hard, the woman could not help boasting again in front of the disguised goddess. This prompted the goddess to warn Arachne that she would have to beg for the forgiveness of Minerva, who would pardon her. An irritated Arachne refused and boasted again, suggesting that the goddess herself try her talent against her own.

Then, Minerva appeared as herself. The people and nymphs who saw this fell on their knees in fear and awe. Arachne, however, did not. Despite being forewarned, she did not beg Minerva's forgiveness. Minerva announced that they should both embroider a picture that would be judged. They set to work. Arachne stitched beautiful pictures of gods and goddesses and their less wise actions, whereas Minerva chose to portray humans being punished by deities. When they finished, Minerva and Arachne looked at each other's handiwork. Arachne realized quickly that she had lost the competition. Although her own work was incredible, that of Minerva was so brilliant that it seemed like it was alive. When Minerva saw Arachne's reckless subject matter, she could see that the foolish young woman had learned nothing from this contest and was not even remotely sorry. Minerva destroyed Arachne's embroidery in anger and then turned Arachne herself into a big spider. Arachne the spider then started to weave a beautiful and delicate web, as she retained her ability to spin and weave. Today her descendants still weave beautiful webs (Fig. 5.7).

The Moon is associated with spinners and weavers of fate. Moon goddesses are often in triad, representing three phases of the celestial body—the full Moon, the first or third quarter and the new Moon. In Greek mythology, the Fates, or Moirae, spin, weave and cut those all-important threads of fate or destiny. Clotho spins the threads, Lachesis catches it up and Atropos cuts it, thereby terminating someone's life. In one Greek myth, there are two jars that contain all the fortunes of humankind. One jar contains all good fortune, while the second contains bad luck. Zeus mixes the contents of both jars up when deciding on a mortal's life. An example of this is King Priam of Troy, a lucky ruler of a famous and prestigious city, but unlucky also, as all of his sons are killed during the Trojan War. The Fates themselves were believed to spin and decide a person's fate when they were born. For the ancients, life was controlled by destiny, but Zeus ensured that fate was adhered to. The Moirae themselves are possibly daughters of the terrifying Night, or Nyx, another link to the Moon.

Because the ancients believed that life was pre-determined, oracles were consulted, the most famous being Apollo's oracle at Delphi. Apollo, among other things, was the god of prophesy. Apollo interfered with the Fates at the wedding of Admetus and Alcestis (the daughter of the King of Thessaly) by getting them

Fig. 5.7 The weaver Arachne's designs are linked to intricate spiderwebs since in legebnd she was transformed into a spider by Arachne. (Illustration by the author)

drunk! Admetus was poisoned by snakes put in the marriage bed by the goddess Artemis, whom Admetus had neglected. When the Fates were sufficiently inebriated, Apollo made them promise that Admetus could live if someone else would die instead of him. As no one came forward, Alcestis, his new bride volunteered. This act of love touched Persephone, the queen of the Underworld, so deeply that she returned Alcestis to Admetus.

The Norns are the Germanic equivalent of the Greek Moirae and as such are connected to the Moon. The Norns warned Odin that the ferocious wolf, Fenrir, would cause his death, which led to his decision to tether the wolf with a magical ribbon, Gleipnir. The Norns were three sisters—Urd, the past; Verdant, the present; and Skuld, the future. Urd's Well was the place where the Norse gods met daily. The three Norns took water from this well daily. This was combined with earth so that the bark from the tree of life, Yggdrasil, did not rot or wear away. The Norns were thought to make decisions about the fate of deities, humans, dwarves and giants alike, when weaving the web of fate. The Anglo-Saxon equivalent of Urd is Wyrd, probably the source of the three weird sisters or witches in Shakespeare's play, *Macbeth*.

Spindle whorls were used by Scandinavian yarn spinners between about A.D. 800 and 1500. They separate the fibers of the wool and can be made from different materials, including oak wood, amber, elk antler, local stone, bone and some metals. Some, including one from post-Viking Greenland, one made from soapstone from the Shetland Isles and the English Saltfleetby spindle whorl, have runes carved onto them that look like stick people and show Viking gods, including Odin and Heimdallr. The Saltfleetby spindle whorl is made from a lead alloy and dates from around the early medieval period in English history. These rune-inscribed spindle whorls were magical objects used by women who worked with wool, to weave people's fate or destiny into wool, like casting a spell. They were most often used to weave in love, luck and protection for loved ones, but they could possibly have been used for weaving curses. This links medieval English and Scandinavian women to the Norns, the Fates and therefore to the Moon itself.

The Moon and weaving feature in the poem by an English poet of the Victorian era, Alfred, Lord Tennyson (1809–1892). The poem is called "The Lady of Shalott." This poem has also been visually immortalized through the paintings of the English pre-Raphaelite artists William Holman Hunt (1827–1910) and John William Waterhouse (1849–1917). In the poem, the people had never seen her. They had just heard singing. The Lady of Shalott was cursed. She could only look at Shalott indirectly, through a mirror, while she weaved her 'magic web' from what she saw in her mirror. The curse became real to her when she looked directly down at Shalott. Then the weaving flew from the window, and the mirror cracked. She is heard at the beginning of the day and at moonlight by the farm workers cutting down the barley crops. The Sun appeared as Lancelot did. Until she discovered the object of her love, Sir Lancelot (a knight from King Arthur's court), the lady of Shalott had only seen life through reflections mirroring the reflected light of the Moon as opposed to the dazzling illumination of the direct light of love from the Sun.

For the Native American Hopi (meaning "people of peace"), Navajo and Keresan tribes, Spider Woman was a creator goddess. Along with the Sun god, Tawa, they created Earth and animals, among other things. The Hopi tribes thought that Spider Woman or Spider Grandmother created the Moon by weaving it from white cotton fabric. The Navajo believe that she taught them to weave. They pay homage to Spider Woman by putting their hands in spider webs before beginning their weaving in the hope that this ritual will help to literally rub off some of the goddess's talent, creativity and cleverness.

In Greek mythology the Moon is represented by three different goddesses: Selene, Aphrodite and Hecate. Selene represented the maiden, the new Moon; Aphrodite the woman, the full Moon; and Hecate the old woman and old Moon. Hecate herself is a dual-purpose goddess. By day she is linked with crop production, but by night she is linked to death, ghosts, spirits and even witches. Hecate is usually depicted with three faces. In the quest of Jason for the Golden Fleece, Medea, the princess of Colchis who helped Jason achieve his quest, called out to Hecate to help with her magic. Hecate is even referred to in Shakespeare's tragedy *Macbeth*. Other more local Greek names for the Moon goddess in the triad are Thetis, Amphitrite and Nereis. Even her offspring represented three phases: Triton, the new Moon; Rhode, the full Moon; and Benthensicyme, the old Moon.

A North American tribal myth, The Moon Wife, explains the phases of the Moon. Two women were in love with the Moon and spent each night lying in a boat made of skin, bathing in the beautiful Moonlight. Their adoration was rewarded, as the Moon himself appeared to them one night in the form of a very attractive man. He announced that the woman who proved herself to be the most patient of the two would be taken up, high into the sky, where she would become the Moon's wife. They were both forewarned that patience was essential, as the work that they had to do up in the sky was difficult. All three went up into the sky, but one of the women felt compelled to look down to Earth. As she did so, she fell down, back into her skin-boat. The remaining woman went to live in the Moon's house as his wife. However, she discovered that life was boring and very, very lonely. She hardly ever saw her husband, as he generally slept all day and worked throughout the night, and she was never allowed to go out with him when he worked.

On a particular night, the Moon let his wife leave their home, but she was told to avoid two particular houses, as they both had a curtain, behind which she obviously must not even peek. Obviously, she ignored the warning. In the first forbidden house behind the curtain were a half-Moon, a quarter-Moon and a tiny Moon sliver, and in the next house were a full Moon, an almost full Moon and another that was a bit more than half full. These Moons were masks, so she was unable to resist trying on one of them. When she did, it got stuck, so she had to confess what she had done to her husband. This all turned out quite well for the Moon-wife. The Moon no longer had to keep the secret from her and even allowed her to help him with being the full Moon and her being the other phases wearing the various different masks.

Another myth that attempts to explain lunar phases is a Native American Indian one. A long, long time ago, the Sky Father had sex with Mother Earth, resulting in the birth of Moon. Moon gradually grew within the stars. When he grew round and full, the Sun dogs started to bite it until it resembled a crescent. This bitten, crescent-shaped Moon still shone up in the sky. Spiritwalker, the leader on Earth of both humans and animals, was worried about the Moon. She got all the animals and people around the square where they danced, hoping that their dancing would make the Moon grow. Then, on a specific night, called the Night When Red Leaves Fall (autumn), everyone on Earth looked up and saw that the Moon was growing again. It grew until it became round, full and content, gradually decreasing, becoming a thin crescent when the Sun dogs bit it again. Then, they all became used to the lunar cycle.

Some mythologies involve heads, faces and the Moon. In an Aztec myth, the Sun was called Huitzilopochtli. His sister, the goddess Coyolxauhqui, killed her mother, Coatlicue, by decapitating her. After this, Huitzilopochtli appeared from Coatlicue's uterus as an adult deity. Then Huitzilopochtli murdered his sister and decapitated her, too. He then flung her head up into the sky, where it became the Moon.

There is an ancient Japanese myth involving the face of a princess and the Moon. A long, long time ago, an omnipotent but compassionate, kind and wise emperor resided on the Moon itself. He looked down on Earth protectively to ensure that all was well. He even had his beloved daughter reside on Earth for a time so that she would learn to love it and its inhabitants, thereby comprehending the importance of protecting the planet from harm. Even today, some still believe that the imperial daughter's face can be seen smiling protectively on the full Moon, looking down on the planet that she had once lived on.

In Japanese mythology, the Moon god was male, Tsukiyomi no Mikoto, the brother of the Sun goddess Amaterasu. Together, they were thought to control the succession of night and day, but they are always pictured sitting back to back. However, Tsukiyomi did not appear to have his sister's good nature. One day, Amaterasu discovered that the goddess of food, Ogetsu, was in the land. She had been there for some time, so the caring Amaterasu sent her brother Tsukiyomi to see if she was all right. He dutifully set off and found Ogetsu, but the journey had made him irritable and ravenously hungry, so he rudely shouted at Ogetsu to provide him with a meal immediately. Ogetsu was upset at his blatant lack of respect for her and made him a meal but decided to include in it vomit and excrement. As she stood looking at the fields, she vomited up rice for him. As she turned to look at the sea, she threw up fish and seaweed. Then she turned around and looked up at the hills, throwing up game creatures.

Tsukiyomi was furious with Ogetsu for serving him a meal produced by vomit and excrement, so he got out his sword and murdered her. From her dead body came all the main Japanese crops, such as rice, millet, soybeans and wheat. Tsukiyomi returned to Amaterasu and told her what he had done. Understandably, she was disgusted with her brother, which is why they are always pictured back to back, so Amaterasu doesn't have to look at her brother. This is also why the Sun and Moon do not appear together in the sky.

The Slavic Moon god is also male in some myths. He is Myesyats, sometimes pictured as a cold balding uncle of the Sun god. However, the Moon deity can be female. In this instance, the female Myesyats is the wife of Dazbog, the Sun god, with a troublesome marriage that includes arguments that are believed to make Earth quake with fear. Myesyats is supposed to have been beautiful, and together the Sun and Moon produced the stars. Their marriage was seasonal, the Sun leaving the Moon each autumn, returning after the cold of the winter and celebrating his return in spring with a re-marriage.

In Bushman mythology, the Moon is regarded as male, too. In a similar story to one previously told in this chapter, part of the longer tale of Kaggen the Mantis and his eland, is a myth about Kaggen and the Moon. The Suricats were unkind to the Mantis, so he found the gall bladder from another eland just lying on the top of a bush. Kaggen took this, made a hole in it and it broke. This created a darkness. Kaggen jumped into this darkness but needed some light, so he removed one of his shoes and flung it up into the sky, where it became the Moon. The Moon is red, as the shoe of Kaggen the Mantis was coated with the red-colored dust from the ground of Bushmanland. The Moon is cold because it is made out of leather. In Bushman astrological mythology, the Sun uses its rays like a knife to cut off pieces of the Moon. This continues until there is only a tiny fragment of the Moon left. Then the Moon apologizes and begs the Sun to stop its torturing for the sake of his children. The Sun does stop, and the Moon gradually grows bigger until it becomes a full Moon, when the Sun's piecing torture starts all over again.

The Sumerian Moon god is also male. He is Nanna, the giver of time. He is believed to be the father of Utu, the Sun and Inanna, Venus. Nanna was Enlil, the creator god's first son. He could be symbolized by horns of a bull as a crescent

Moon. Nanna was then regarded as the cosmic bull in charge of the stars, which were his celestial cattle. The Babylonian equivalent of Nanna was Sin, the god of the Moon and the measurer of time. His symbol was also the crescent, and his children included Shamash and Ishtar. The number associated with both Nanna and Sin was 30, which represented the portion of time in a month—30 days.

Like many other Moon deities, Nanna sailed across the sky in a boat in the shape of a half-Moon with a sail on it. The full Moon was his crown. He was supposed to be able to calculate time by his waxing and waning. He is also believed to have been able to use his light to show up the evil desires and deeds of those with less than honorable intentions.

Sin, the Babylonian/Assyrian Moon god, is linked to the Polynesian Moon goddess Mahina, Hina or Masina. Polynesia can include the Hawaiian Islands, Samoa, New Zealand, Tonga, the Hervey Islands and the Fate Islands, among others. They each have different traditions and mythologies regarding the Moon. Hina is a major Hawaiian ancestral goddess of the Moon. Her name means "girl." She is a very complex goddess with many Kinolau—bodies or personae. She can be 'Hine,' ea, the goddess of the sunrise and sunset and Hina,' i ka malama, who is the Moon's foster daughter, among many others. In a Hawaiian myth, Hina resided on Earth, living the life of a mortal woman. She did physically demanding work each day, creating plaited mats for the family to sleep on from Hala tree leaves. She also made torches from Kukui tree nuts in addition to food preparation and cooking. Her family did not make her happy, either. Her children were disorderly and her husband was idle and not resourceful.

Eventually she decided that she had had enough of living on Earth and made up her mind to escape by climbing up the rainbow through the clouds to the Sun, which was shining especially brightly that day. She left her home very early in the morning, when her family members were all fast asleep. As she climbed up the rainbow, the extremely high temperature of the Sun's rays glaring fiercely down on her sapped some of her strength. Still she continued her upward journey towards the Sun until its extreme heat became too much. Then she slid downwards, down through the clouds, down the rainbow and back down to Earth. Her strength gradually returned as the day passed. Then she saw the bright, full Moon up in the sky and decided to climb up to it so that she could escape from her physically difficult and miserable life here on Earth.

As she started to climb up, her husband saw her and asked her not to go, but she told him that she'd made up her mind and that she was going to the Moon, who would be her new husband. She continued climbing, but her husband leaped up to her and grabbed hold of her foot. She tried to continue climbing and managed to stop him holding onto her, but as he fell down to Earth, he refused to let go, pulling off her lower leg. Hina continued to climb up until she eventually arrived at the Moon.The ancient Hawaiians believed that she could be seen up in the Moon and has been given the name Lono-moku by the Hawaiian people, meaning the "crippled Lono."

Hina has many different aspects. She reigns over tides and has shown women how to produce fabric from tapa bark. As Hina-Ika, she is believed to have thought of making fishing nets that were rather beautifully woven from her long, silvery lunar hair.

The Hervey islanders have a different story. They believe that the Moon or Marama had looked down on Earth and seen the beautiful Hina. He liked the look of her and persuaded her to go and live with him.

To the Torigan islanders, Hina is still in the Moon, becoming the goddess of the 'fire-walkers.' In New Zealand, they call nights when the Moon is not visible 'dark Hina.'

For the Samoans, Hina and a child were accepted by the Moon complete with tools and materials to make tapa cloth. They believed that Hina made clouds out of tapa and looked for her during a full Moon.

In Mesopotamian mythology, Anzu's symbol was a sickle Moon with a star and a tree. Despite her symbolic representation, including a Moon, she was not a Moon goddess but a goddess of chaos, who was sacrificed to create the universe itself.

Anunitu/Annunita was a Babylonian/Chaldean goddess of the Moon. Her symbol was a disc with eight rays. She later became Ishtar. Astarte is the most important female Phoenician goddess, connected to other Moon goddesses, including Ishtar, Selene, Aphrodite and Artemis. Zarpandit was an early birth goddess in Assyrian and Babylonian mythology and was worshipped every night when the Moon rose up.

Thoth was the Egyptian Moon god, also god of words and of knowledge itself. He is believed to have come from one of the eyes of Horus, the sky god. He ended up taking over from Horus, reigning over Egypt for three millennia. Then he rose up into the sky, becoming the Moon. He is usually shown as an ibis bird, whose beak resembles a sickle Moon. In one myth, he was commanded to light the sky by Ra, the Sun god. Then he was slowly consumed by creatures that were made to vomit him up.

Thoth, being a god of words, knowledge and writing was the obvious choice of deity to record the judgment of the souls of the dead. Thoth merely records this—he does not make the actual judgment. Thoth appears in a myth involving other Egyptian deities. Nut, the sky goddess, married her twin brother, Geb, the god of Earth, despite being commanded not to by Ra, the Sun god. Ra was furious at the 'siblings' disobedience and got Shu, the god of the air, to force them apart. He pulled Nut up, where she became the firmament or sky and pushed Geb down, where he became Earth. Nut was given the further punishment of not being able to have children in any day of the year. Thoth decided to help her. Thoth and the Moon had a game of draughts. Thoth won and claimed light as his prize. He won 5 days' worth of light, so was able to make an extra 5 days. These days were when Nut gave birth to her children Osiris, Seth, Horus, Nephthys and Isis.

Other Moon gods include Wadd, from southern Arabia. Wadd is even mentioned in the Qur'an as a pagan god. Northern Arabia had another Moon god, Aglibol, who was pictured with a crescent Moon on his body. The Phrygian Moon god was Men, who also commanded the sky and the Underworld. (Phrygia was part of what we now call Turkey.)

Fig. 5.8 The reflection of the Moon is not to be disturbed in Chinese mythology. (Illustration by the author)

Ancient Greeks worshipped Io, a Moon goddess who was symbolized by small horns. The cow was linked to the Moon because of the similarity between the crescent Moon and the shape of cows' horns.

In Chinese mythology, elephants upset the Moon goddess. A long, long time ago, there was a lengthy drought. A herd of elephants had been traveling for days in search of water. They were about to give up when they discovered an entire lake of beautiful, clear water, so they ran towards it, stampeding over hares in the process. Since the hare is a lunar creature, associated with Moon deities, this was not a wise thing to do. The Moon goddess was also angry with the unfortunate elephants for getting in the way of her beautiful reflection by drinking the water in her Moon lake (Fig. 5.8). As soon as they realized that they had upset the Moon goddess, they took one last drink and swiftly left, taking care not to step on any more hares.

The Moon seems to be an actual person, a man, in some mythologies. In an Oceanic dreamtime myth, the Moon was a young man who had a sister called Dugong. One day, they decided to dig. While they were both busy with their digging, Dugong got bitten and jumped into the sea to cool the pain. She transformed herself before doing this into a Sirenian of the Asian seas, now called a Dugong. Moon did not like being alone, so he jumped into the sea, too. Dugong died, but Moon did not, with his bones becoming sea shells. Moon was rescued by eating eat lotus and lily bulbs.

In a North American Pawnee myth, Pah is the Moon god, with his wife being the Sun goddess Shakaru. The Moon is male, too, in North American Inuit mythology. He is called Aningan or Igaluk and is believed to have his own igloo up in the sky. As young children, Aningan and his sister, Malina, were happy and loved to play together. They usually played in the dark, even as adults. One night, Aningan became envious of his sister's lovely radiance and started a fight with her. Malina had been holding a lamp made from lamp moss and seal-oil when Aningan had come up to her and started the fight, and Malina's fingers became blackened from it. They continued their fight and she touched his face, making it all black.

Having a dirty face infuriated Aningan, so Malina sped off as far away from her brother as was possible. This happened to be up, up high into the sky, where, with his dirty face, he became the Moon. This never-ending chase explains why the Sun and the Moon are not in the sky at the same time. The constant chasing is hungry work, so for 3 days each month the Moon vanishes to allow him to hunt and to eat enough to continue the chase. Aningan is supposed to live with his demonic cousin in the igloo. Both Ainigan and the cousin, Irdlirvirisissong, love to hunt. Although the Moon is by far the more successful hunter, Irdlirvirissong loves to venture into the sky, where he can dance around and make everyone laugh. However he is someone to fear, because if he catches any people around him, he will eat their intestines!

In Norse mythology, Mani was the male personification of the Moon. He had a brother, Sol, who is the physical manifestation of the Sun.

However, in another myth, the Moon was such a devoted son that he insisted on following his mother around absolutely everywhere. This irritated the mother. She was so exasperated with her Moon-son that she flung her dishcloth in his face, an explanation for why the Moon's face is dirty.

Other Moon deities are female personifications of the Moon. Lona in Hawaiian mythology was a Moon goddess who, perhaps unwisely, fell passionately in love with a man by the name of Aikanaka. They lived happily ever after, at least for many years. They were finally separated by the death of the mortal. This is quite unusual for a goddess.

In Chinese mythology, the Moon goddess, Heng-o, behaved in a very maternal, caring way towards her children. A very long time ago, the Chinese thought that there were 12 Moons, one for each month of the year. (They also believed that there were ten suns, one for each of the ten days of the week.) As a caring mother, Heng-o took her children to the West and bathed each of her Moon children. Then, one Moon child would journey in a special chariot for a lunar month until they reached their destination, the East.

Shelardi is an Armenian Moon goddess, similar to the Babylonian Moon god Sin, although she is the sister of the Sun. Kilya is the Inca Sun god, Inti's wife and a lunar goddess. She is also known by the name Mairna Kilya. She has a shrine dedicated to her inside the Sun temple at Cuzco. It has walls coated with silver. Silver was regarded as a sacred metal to the Incas of Peru. It was supposed to be the tears of Kilya. The Moon, Kilya, was used by the Incas to determine when their religious festivals, rites and celebrations were to be held.

The Moon is sometimes credited with having some human attributes or emotions. This is the case in the Maori myth of Rona and the Moon. Rona was the sea god Tangaroa's daughter, who was believed to control the tides. At night, she went to a nearby stream to fill up her bucket with drinking water for her family. On her way back, it suddenly got very, very dark. The Moon had hidden itself behind some clouds, and Rona could not see in this darkness. She continued to walk, but she tripped up. She was annoyed with herself for not being able to see where she was going, and so blamed and cursed the Moon horribly in a rather too-loud voice. The Moon heard these horrible loud curses and, in turn, cursed the Maori people, feeling hurt and angry—very human emotions. He also picked up Rona and her bucket. Some people believe that they can see a woman with a bucket up in the Moon. The Maori people believe that it rains when Rona knocks her bucket over. Confusingly, in another Maori myth, Rona is the name of a man. Rona's wife left him, and he searched high and low for her, traveling for miles and miles, eventually being reunited with her on the Moon. The Moon's phases were explained by Rona and his wife taking turns in eating each other, becoming gradually thinner; then they renew themselves in the waters before starting their devouring of each other all over again.

The Moon appears as a sky and Moon god, Arebati, a father figure believed to have made men and women from different kinds of clay, using red clay to create the African Mbuti people who are also known as the Pygmy people of the Congo.

The Egyptian god Khonsu's role was similar to that of the African Arebati. He was the god of both the Moon and of youthfulness. He is usually depicted as a falcon. He was the son of Amin and Mut, deities associated with the Sun. Khonsu means "traveler," probably connected to the Moon's voyage across the sky each night. Like Arebati, Khonsu is believed to have been a creator of new life, somewhat of a caring paternal figure.

However, the Slavic Moon god Myesats, in mythology, was thought to have very lustful human desires. Perun/Perkuna, the chief deity and thunder god (who is also associated with fire, war and weapons), was furious with the behavior of Myesats and decided that he needed to be taught a lesson. Myesats was married but conveniently forgot about his wife, the Sun, when he made sexual advances to the Morning Star. Perun stabbed the Moon angrily, cutting him up as punishment for his inappropriate behavior.

Sometimes gods and goddesses are divine representations of the Moon, like Luna in Roman mythology, whose Greek equivalent is Selene. Luna is a part of the triple Moon goddess together with Hecate and Proserpina, each goddess representing a different lunar phase. The ancient Columbian goddess, Chia, was also a triple lunar goddess. The Sami Moon goddess was Mano. The people honored her around the time of the new Moon by being extremely quiet, especially around the time that we now call Christmas—not an easy task. Mano was treated with great caution and respect, as she was inclined towards risky and unstable behavior.

Some stories merely involve the reflection of the Moon rather than the actual Moon and the stupidity of others. In a story from Turkey a young man by the name of Hodja set off to the well to collect some water. He looked into the well in order to complete his task and saw the reflection of the Moon in the water. His first thoughts were that the poor Moon had somehow fallen into the well and needed to be rescued. He found a rope that had a hook attached to it nearby and threw it down the well. Unfortunately for Hodja, the hook caught, causing the rope to snap. Hodja holding onto the rope at the top of the well fell backwards when this happened. Then, Hodja noticed that the Moon was back where it belonged, up in the sky. He did not realize that the Moon had never been in the well.

In an Indian Tibetan story were a group of monkeys. One day, the monkeys also saw the Moon's reflection in a well. Like Hodja, they all thought that the Moon had fallen into the well. The leader of the monkeys asked the others if they should rescue the poor Moon, and they all agreed that they should try to help. They had the idea of making a chain by holding onto each other's tails in order to reach the Moon. The first monkey held onto a nearby tree branch, with the other monkeys holding onto the tail of the previous one. The branch buckled under the weight of all of the monkeys. The wind started to blow. This disturbed the water, and the Moon's reflection seemed to vanish. The branch snapped, and all of the monkeys fell into the well.

In Finnish mythology, a teal, which is a member of the duck family, laid an egg upon the lap of the air's daughter, Ilmatar. Ilmatar floated upwards and the egg fell down. The egg white was then transformed into the Moon.

Chapter 6

Moon Madness, Superstition, and Other Lunar Associations

Although some creatures fear the Sun's very bright light, many thrive in the more gentle reflected light of the Moon, like dwarves, giants and even fairies. In old English mythology, the Asrais liked to bask in the light of the Moon, as they needed Moonlight to grow bigger. (Asrai were a kind of fairy that looked like mermaids with fishtails rather than legs and green hair that looked like seaweed.)

Other, more sinister creatures thrive or even come to life in the reflected light of the Moon, including demons, ghosts and even vampires. Perhaps the concept of blood-sucking demons came from vampire bats found in South America. They have razor-sharp teeth. Their saliva contains anti-coagulants, which cause their victim's blood to flow and flow and flow. They hunt for blood under cover of darkness, as they can see very well in the dark. They like to torment their victims before biting them in places that are difficult to defend, like the rear end! Vampire bats live in dark caves in the daytime and feed entirely on blood from other animals, especially large mammals. Although they are small creatures themselves, they can consume far more than their own body weight in blood and are able to regurgitate blood to share with their young and other hungry bats in their colony. This sounds uncannily similar to Dracula mythology!

The Keres obeyed the orders of the Fates, who are associated with the Moon as spinners and weavers of destiny. They lingered at battle sites, rather like the Norse Valkyries. Unlike the Valkyries, they had claws and wore red, the color of the blood that they drank from their allocated victims.

The Empusa or Mormolykiai, which is what Lamia became, were demonic. They are evil vampire-like spirits who enjoy preying upon sleeping victims. They are frequently able to change themselves into humans, especially beautiful women,

© Springer Science+Business Media New York 2015
R. Alexander, *Myths, Symbols and Legends of Solar System Bodies*, The Patrick Moore Practical Astronomy Series, DOI 10.1007/978-1-4614-7067-0_6

and devour their prey by moonlight or in darkness. The Empusa are commanded by Hecate, one of the Greek Moon goddesses.

In Japanese mythology are a variety of demonic creatures, including the Kokiteno, who appeared as beautiful foxes to tempt gullible men. These demonic foxlike ghostly creatures are associated with the Moon, perhaps because foxes are nocturnal creatures. The fox's alluring red-brown fur links it to the fires of hell and the color of blood. In ancient China powder from ground-up fox testicles mixed with wine was believed to act as an aphrodisiac. The ancient Chinese also believed that foxes lived for a millennium. Foxes could be the steed of choice of other spirits to mount and ride upon. The god of rice, Inari, was one such spirit. The Kokiteno could also transform themselves into beautiful women and are also known as werefoxes, connecting them to the Moon, like werewolves. They were also known as fox-women, with uncontrolled sexual desire that could take away the life, vital spirit or soul of a male victim, like a vampire.

Slav people believed that vampires fed on blood to preserve their own 'life.' However, there are other types of European vampires. The Sampiro from Albania tend to be Turkish Albanians who like to spread doom and havoc wherever it wanders by moonlight, wearing grave clothes. The Serbian Vlkodlak are people who have killed, lied under oath and had sex with their siblings. They might also have eaten creatures that werewolves have slain or themselves have been the victims of a werewolf. The Norferat from Romania allegedly can cause married men to become impotent. They were believed to have become vampires by unfortunate birth circumstances, including being the seventh son of a seventh son or by being the bastard of bastard parents. In local Transylvanian folklore, it is believed that the devil and his followers are supposed to live on the peaks of the Carpathian Mountains.

The most famous vampire of all is Dracula, immortalized in literature by the Irish author, Bram Stoker (1847–1912). This Victorian work of fiction was first published in 1897, drawn from pre-existing vampire mythology. However, the novel became the basis and the inspiration for countless subsequent movies, television programs and novels, including more recent teenage vampire stories. Stoker's protagonist is based upon a genuine historical character, Vlad the Impaler, who lived in the fifteenth century, so called as he enjoyed impaling his enemies. This figure was also called Dracula, which can mean "devil." However, the historical character did not claim supernatural powers or to be in any way immortal. He died on the battlefield, and his severed head was given to the Sultan of Constantinople.

A vampire is supposed to be a spirit or dead body brought back to some sort of 'life' at night by sucking the blood of sleeping humans, thereby transferring the life blood or life force from the living to the 'undead.' *The Undead* was allegedly the original title for this Gothic horror novel. *Dracula* is written in the form of journals, letters, diaries and memoranda of the main characters, but not of the protagonist, Dracula, although he did impart information to one of the main characters, Jonathan Harker. In this novel, Dracula first appears in the form of a man. Jonathan visited Dracula; he was kept in his Transylvanian castle against his will but did manage to escape. Dracula, in a coffin, traveled to England, where he sucked the life-blood

Fig. 6.1 Garlic flowers supposedly ward off vampires, in myth. (Illustration by the author)

from various people, including a woman, Lucy Westenra, who became a vampire herself before her corpse was decapitated and a stake hammered through her heart. The novel had allegorical significance, exploring themes such as light versus dark, Christianity versus darkness, the devil and the undead.

From this novel, we learn much about vampires, drawn from existing superstition. Vampires are repelled by consecrated wafers and crucifixes, garlic, mountain ash and wild roses, used in old herbal remedies (Fig. 6.1). Van Helsing, the Dutch hero, placed a wreath of garlic flowers around Lucy Westernra's neck and around all points of entry to the room—the fireplace, window sashes and door jambs. A central part of the plot involved the destruction of all 50 boxes of Transylvanian earth that had traveled with Dracula to England, denying Dracula his daytime resting place. Van Helsing and his assistants destroyed each box of earth by re-sanctifying it to God. This became a race against time, eventually won by Van Helsing. They intercepted Dracula's body on its way back to Transylvania, where his head was severed and his heart was pierced. Then his entire body disintegrated.

Dracula and the creatures under his command seemed to thrive by the light of the Moon. The wolves howled in the moonlight, ghostly beings come to life from moonbeams, and Jonathan Harker's fear (see below) vanished when the Sun rose. The patient in the mental asylum, Renfield, calmed down when the Moon rose until sunrise, associating the Moon with lunacy. The gray wolf escaped from the zoo just before midnight, and at full moonlight Dracula, in the guise of a bat, became emboldened and tried to smash the window to Lucy's bedroom by hitting it with its wings. Vampires can appear upon rays of moonlight, and Dracula could command some of the elements—thunder, storms and fog. Dracula could also command most nocturnal creatures, including wolves, foxes, bats, owls, rats and moths (Fig. 6.2).

Vampires have no shadows, as noticed by Bram Stoker's character Jonathan Harker in a dream that seemed like reality. Shadows can symbolize a person's soul, so vampires are soulless.

Fig. 6.2 Bats and vampires both have associations with the Moon. (Illustration by the author)

Dracula, in folklore, is somewhat Pied Piper-like, stealing the life-force in the form of blood rather than children. Blood is the very stuff of life, symbolic in the Christian ritual of Communion, where the wine that is consumed symbolizes the blood of Christ, shed to save sinful humans. The wafer is sometimes used in similar rituals or bread is used, representing the body of Christ, which is why Dracula promises the asylum patient Renfield many lives if he would worship him, just like Saturn trying to tempt Christ in the wilderness. Dracula, Satan, the night and therefore the Moon are associated with spiritual darkness as opposed to the spiritual illumination of Christ and the Sun.

Bram Stoker's Dracula did not like mirrors. There were no mirrors in his Transylvanian castle, and the character Jonathan noticed that Dracula had no reflection in his own mirror. Dracula's reaction to Jonathan's mirror was to purposely shatter it. The mirror can be symbolic of the soul, which would explain Dracula's lack of reflection and also the clear surface of divine truth as opposed to dark, Satanic untruths or lies.

A mirror also represents self-realization. Dracula would have been forced to confront his own truth, if he had seen his own reflection, which explains why he broke Jonathan's mirror. Dracula would have undoubtedly died if he had seen his own reflection or himself as seen by others. The mirror is finally a symbol of the Moon, as it also reflects light. Some believe that mirrors can absorb emotions, making them potentially dangerous. The ancients believed that people had some kind of enchanted connection with their own reflection and that a mirror had the power to keep souls. This explains why mirrors were covered or made to face the wall in a room where someone has died or was laid out. Otherwise, the dead person's soul could remain trapped in that room or in the mirror for eternity, never reaching any kind of afterlife.

Some Chinese believe that glass and therefore mirrors can be a doorway to the afterlife. Therefore, a mirror is an entrance for evil spirits to enter this world and need to be covered over, especially at night, when darkness allows evil spirits more freedom.

In ancient times, some people in Britain believed that a mirror was a doorway to the next world. Therefore, people could walk into or, more worryingly, be dragged into another world through a mirror, where they would remain trapped. Mirrors made from obsidian or volcanic glass were found in the Neolithic town of Catalhoyuk (Turkey), revealing that people have been fascinated by reflective surfaces for millennia. It is conjectured that these mirrors might have belonged to priestesses or perhaps they served as magical objects that were used to see into the future. In East Yorkshire, England, a rare Iron Age chariot burial was discovered in 2001 of a local woman of around 40 years old. She was buried with an iron mirror, perhaps for looking into the next world? (A mirror would have been one of two ways that Iron Age people would have seen a reflection; the other would be from looking into still water.)

According to the principles of the ancient art of feng shui, mirrors create a strong form of energy, or chi, not conducive to a decent night's sleep. It is therefore recommended that mirrors be covered up at night to help with relaxation and sleep. There is an old Chinese myth that offers another explanation of why mirrors are covered at night in China. A woman had not looked in a mirror or even glimpsed her reflection in water for decades. Then, one day, she decided to look into a mirror. She saw an ancient, wrinkled old woman and refused to believe that it was her, as her memories of herself were as being youthful, fresh and incredibility attractive. She put the old woman reflection down to an evil demon trying to change her. As this old woman was part of the Chinese ruling family, they decided to humor her by covering all mirrors and reflective surfaces at night.

Mirrors can also be used as protective charms against evil spirits, who have no reflection, like Dracula.

In Japanese myth, a prostitute known as the Lady from Hell, or Jigokudayu, looked into her mirror as usual. However, instead of seeing her own beautiful reflection, she saw a skeleton looking directly at her. This was a life-changing experience, or enlightenment, for the woman, and she became one of the followers of the Zen master Ikkyu Sojun, also known as the holy mad or crazy man, as he danced publicly with a skull on the top of a pole and was known to visit places of ill repute.

In Japan, the mirror is one of the imperial treasures and an object appropriate to the Sun goddess, Amaterasu. A holy mirror is temporarily given to each new emperor, symbolic of his official position and linking him to his royal and divine ancestors.

The mirror can be associated in Christianity with the Virgin Mary, because she carried and gave birth to Jesus, God's son and a reflection of himself. The Moon is also associated with Mary. The Moon reflects the Sun's light, and she gave the world God's son, his reflection as well as herself being illuminated from her son's holy light.

Reflections, first seen in lakes, rivers and pools of water by the ancients, eventually led to the belief that a reflection was literally the soul in physical form. Therefore, if a reflection is harmed in any way, the person that it belongs to will suffer, too.

Mirrors, similar to crystal balls, have been regarded as magical, enchanted objects that can foretell the future, which is referred to as scrying, perhaps the most famous scryer being the English doctor John Dee (1527–1608).

Witches' balls, reflective spheres of glass, were believed to offer protection against witches and spells, the idea being that these objects were attributed with being able to bend the evil eye's influence away from them.

According to superstition, a young girl can tell the number of years that she has to wait until her wedding by venturing outdoors carrying a mirror on a night when there is a full Moon. She would then have to stand upon a stone while still holding the mirror with the Moon behind her. When she looks at the Moon through the mirror, she might see smaller Moons. If she counts those Moons, it would reveal how many years before her wedding would occur.

In Ghanaian superstition, mirrors are very dangerous indeed, especially when lightning is added. The reflection of lightning alone is enough to kill someone, and during a storm, behind your own reflection can be seen the faces of your enemies.

Wolves are associated with the Moon, probably mainly because they are nocturnal creatures that do sometimes howl at the Moon. They are also portrayed as familiars of witches, warlocks and even of Dracula. Paradoxically, wolves could even represent the morning Sun, as they can see in the dark, which also has very positive spiritual implications. Despite this, the wolf tends to have acquired a bad reputation in fairytales and in mythology. There are many, many wolves featured in mythology and folklore.

It is claimed that Lycaon was the first werewolf. In Greek mythology, Lycaon angered the chief god, Zeus, who, in retaliation, changed him into a wolf. Zeus also, as punishment for an act of hubris, transformed all but one of his sons into wolves, too. Technically, they all became 'man wolves' or 'werewolves,' although they remained as wolves, not changing at the full Moon.

Native American culture is steeped with tales involving the wolf, which is associated with Sirius, the dog star, itself a guide in the sky and believed to be connected to the Moon and to have special powers. Wolves can also become the mounts of witches and wizards. In a Siberian myth, however, it was believed that female demons could transform themselves into wolves, and the Mongolian hero and conqueror Genghis Khan claimed to have descended from the gray wolf and even from the sky itself in some myths.

Some of the supposed werewolves were people who most probably suffered from the real medical condition of lycanthropy. People suffering from this condition believe that they are wolves and acquire lupine characteristics such as moving on all four limbs. This is what is believed to have happened to the powerful and very rich king of Babylon, Nebuchadnezzar, in the Bible. He appeared in the book of Daniel and put Shadrach, Meshach and Abednego into the fire and Daniel into the lion's den. Daniel's God cursed Nebuchadnezzar and turned him into a beast of the field, as he did not repent, striking him down with lycanthropy.

In Germany, towards the end of the sixteenth century, a man by the name of Peter Stubb was supposed to have confessed that he was a vampire. He said that he changed into a wolf when he wore a belt made from the skin of a wolf. He was supposed to have raped, savaged, murdered and finally eaten many young women. He was also accused of murdering his own son in the local woods. Peter's confession was extracted by torture, and so is therefore unreliable. It seems like the locals used the myth of the werewolf to try to make sense of the horribly savage deaths of some people. The unfortunate Peter suffered a horribly gruesome death.

This is reminiscent of the fairytale "Little Red Riding Hood," or is even more like the version of the story "The False Grandmother," where Red, the female protagonist, is a woman of very loose morals but gets eaten by the big, bad wolf. The brothers Grimm cleaned up the original bawdy story to make it more child-friendly. However, in all versions of the story, the wolf is a personification of evil, cunning and treachery. There are many British and European werewolf myths. One of the most frequently told involved someone harming a wolf in the night, and the next day, a person appears with exactly the same wounds.

Wolves have always been considered to be dangerous creatures capable of attacking humans. In medieval bestiaries, the wolf is a devilish animal. It is compared to Satan himself because the female wolf's eyes glow at night so brightly as to stun human senses, like Satan's blinding stare. In ancient times, the wolf was regarded as a spectral creature with a look that could rob a man of the power of speech.

The final British wolf was killed in the eighteenth century. In the fifteenth century Scotland's King James I wanted the wolf population annihilated, but this did not happen quickly, and it is believed that the final one was murdered in 1743.

During the reign of the English King Edward the Confessor (1042–1066), convicted criminals were made to wear a wolf-headed mask, turning them into wolfmen and reducing them to savage, feared and unpredictable beasts.

It was believed hundreds of years ago that people could voluntary become werewolves. In order to do this people had to make a bargain with Satan himself. The person's soul was swapped for the seductive power and strength of the werewolf. Wolf-skin belts were symbolic of this and were supposedly required to transform into a werewolf. Some Viking warriors chose to wear the skin of a wolf and drank the creature's blood to try to evoke the wolf's strength, savagery and boldness on the battlefield. Again, they were literally man-wolves.

Werewolf sightings first began in the European countryside, but nearer to towns in the fourteenth century. Werewolves were believed to have been people who had the ability, whether voluntary or not, to change into ferocious, uncontrollable beasts at night usually during a full Moon. They were supposedly immune to all weapons, with the exception of those that were made of silver or silver bullets. The silver bullet or other weapon was required to penetrate the creature's heart, at which point the wolf would transform back into being human. Silver was key because it was associated with the Moon and because of its connection with purity.

The Moon is often referred to as 'silvery' or as giving off a silvery glow. This could date back to ancient Greek times, when each of the known celestial orbs (including the Sun and Moon) were allocated with a metal, and later, a color.

The Moon's both allocated color and metal was silver. The ancient Greeks were right. In 2010, NASA scientists revealed to the world that Moon dust contained water and particles of silver, so the Moon really is silvery!

Silver has been used since ancient times. The ancient Greeks used silver to make coins, for example. It is now used for many different things. Pre-digital photography used paper with silver salt on it for the printing of images. Silver was used for elaborate, status-symbol tableware in the days when people had servants to keep it from tarnishing, and is now used for jewelry. Silver can be used in cosmetics and is even available in a bar of expensive soap containing nanoparticles of silver, giving it antibacterial qualities, interfering on a molecular level.

Silver is regarded as a lunar metal, cold and feminine and shiny. It is obviously an object or substance appropriate to lunar goddesses and female royalty. It shines with a whitish glow, which is symbolic of virginity, purity and hopefulness. In language, the metal silver is a noun, whereas silver the color is both an adjective and a symbol or concept.

Silver is used frequently in the English language. The Silver Age is a less prestigious period than the Golden Age and is believed to have been after the Golden Age but before the Iron Age, so silver is often second best, as in the order of medals at the Olympic Games. The silver jubilee, or anniversary, celebrates 25 years, while the golden one represents a much more impressive 50 years.

To be born with a silver spoon in your mouth means that you are born into privilege, comfort and wealth. To have a silver tongue means that you always know what to say to please people or that you are skilled at telling people what they want to hear, while every cloud has a silver lining means that something good comes out of something bad. Hair that is graying can be described as silvery, as can the beautiful Moon itself and the sound of pealing bells. A silver birch is a type of tree with a whitish bark that looks silvery at night. A silver fir tree looks silver in color, as does the silver fish, either a type of goldfish or a fish-shaped insect. The silver screen is the cinema or movie screen. Thirty pieces of silver is the price for which Judas Iscariot, one of Jesus' disciples, betrayed him.

Silver—because of its association with the Moon, a celestial body and one of God's creations and because of its symbolic purity—has been used in the making of icons and for Bible covers. The Italian painter and architect Giotto di Bondone (1267–1337) used silver leaf in his painting of the apostles, "The Last Supper" (1320–25). Artists used gold on Jesus' halo, to emphasis his divine status. Silver inevitably tarnishes. Some old Bibles had solid silver covers that would have been very expensive and became family heirlooms, emphasizing the spiritual importance of this book. Bible covers tend to have images of the Sun and Moon on them and sometimes skeletons or skulls, emphasizing Jesus' triumph over death. Some exquisitely painted icons were partially covered with solid silver, once again to emphasize spiritual purity.

The pearl is the feminine, yin, watery principle, symbolizing virginity, purity and the Moon. Pearls have been referred to as the teardrops of the Moon since ancient times. Crystals are associated with the Moon, as they reflect light, just like the celestial body itself. They are also receptive like the Moon. The crystal ball, however, is usually made from clear glass, but is an orb, like the Moon, and symbolizes perfection and light.

The Moon is associated with madness. The word lunatic is derived from the word lunar, meaning "related to the Moon." An ancient superstition suggests that people who stare too long at the full Moon will become lunatics, growing crazier at each subsequent full Moon. The behavior of animals is supposed to become stranger and more unpredictable at the full Moon, linking them to the mythical werewolf. Witches and demons such as Dracula seem to be able to draw power from the full Moon for their own evil purposes. There is supposed to be a higher crime rate at the time of the full Moon, known as the Transylvanian effect, although there is as much evidence to disprove as there is to prove this. Pliny the Elder (A.D. 23–79), the Roman officer and author, believed that the brain was the most moist part of the human body, and therefore the most likely to be affected by the Moon, which rules the tides. However, the gravitational pull of the Moon only seems to affect water such as seas and oceans, not enclosed water like the human brain or body, which is predominately water.

There are links between darkness, loneliness, madness and the Moon. There are many stories of people going mad in the Antarctic, where the Sun sets in March at the South Pole and will not rise again for 6 months. In the Arctic, there are stories of men going mad from the darkness and loneliness, who end up either killing themselves or others. It seems that for some people, the light of the Moon is not enough.

Some tribes and people in the Arctic are dependent on the light of the Moon (the more subdued, gentle, reflected light from the Sun) for the months of the year where there is no Sun. The Inuit people regard the Moon as the sender of snow.

The Dolgan tribes are reindeer people. They drive the reindeer across the Siberian tundra in search of food. They insulate their huts with reindeer fur, from which they also craft clothing. These people live on raw fish and would never eat a reindeer unless it was a matter of life and death. The reindeer is a symbol of the Moon (Fig. 6.3). Tribes such as the Dolgan are dependent upon both the Moon and the reindeer for their very survival in extremely bleak, bitterly cold, subzero temperatures. The reindeer can also be related to night and the realm of the dead as a guide of souls. In Siberia and Alaska, bears were associated with the Moon, since they are animals that hibernate; they come and go like the Moon itself.

To the Coast Salish peoples, the Moon is a transformer. His parents were a mortal woman and a star. The Moon is supposed to have given the people salmon and other fish and game. However, to the Maidu, the Moon is a less benevolent power that steals children. The Moon is associated with magical powers for Shamans. Alinnaq, the major deity in the western Arctic, is associated with the Moon. Shamans ask for help from Alinnaq personally when there is very little available food. The whale hunters ask Alinnaq to ensure that they have a safe and successful hunting trip in spring. This involves a ritual culminating with the women standing on their igloos as the Moon rises. Then they shout and lift water that had been pre-blessed by the Shaman towards the hole in the sky that was thought to connect Earth to the Moon. The idea was that if this holy water got to Alinnaq, this would

Fig. 6.3 In the Arctic, sacred reindeer symbolize the Moon to the Dolgan people. (Illustration by the author)

please him, and, in return, he would drop rough lamp-tar images of whales into their pots. These crude images then would became potent hunting amulets. The major Moon spirit of the Inuit people was Tarqaq, who was also associated with fertility and animal spirits. He also had the power to call forth bad weather and unsuccessful hunting trips that could result in starvation and illness as punishment for immorality.

The crescent Moon is a universally well-known symbol of Islam, and together with the star appears on the flags of many Muslim nations, including Pakistan, Turkey, Algeria and Tunisia. The crescent Moon is symbolic of divine power and influence, growth, rebirth and resurrection and, with a star, represents paradise. The red crescent is also internationally recognized as the equivalent to the Red Cross, both symbols representing a humanitarian movement that exists in almost every country.

The crescent symbol itself originally symbolized the Greco-Roman goddess Artemis/Diana. It was then used by the city of Byzantium (now Istanbul) as the city's symbol. This occurred before Jesus was even born. This symbol and flag was then used by the Turks when they took over the city of Byzantium/Constantinople in 1453. According to legend, Osman, the sultan and the man credited with founding the Ottoman Empire, is supposed to have had either a vision or a dream where a crescent-shaped Moon kept on growing until its 'horns' or ends stretched from east to west. He regarded this as an auspicious sign and adopted it as his dynastic symbol. However, this symbol is itself somewhat controversial, as Islam historically was without a symbol, and some do not want to accept a pagan sign to represent their religious faith.

The Islamic faith spread from Arabia to Africa and Europe in the West (Egypt, Morocco, Alegeria, Libya, Russia, Finland, Turkey, Greece, Balkan countries, Spain, the south of France, Romania, Portugal, then stopped at the gates of Vienna) in addition to, in the East, India, Malaysia, China and Indonesia. If a line were to be drawn from east to west, joining the countries where Islam spread, it would resemble a crescent or waning Moon in the west. The star on flags is symbolic of a shining light, or glad tidings, as the Star of Bethlehem was symbolic of the coming of Jesus, the son of God.

It is interesting that the new regime of Libya changed their flag from the plain green field of Gaddafi to the flag of the rebel army, which has three horizontal stripes; the top is red, the bottom is green and the middle is black, containing the crescent Moon and the star, the emblem of Islam and of paradise.

The Republic of Palau, consisting of 26 western Pacific islands and 300 islets, has a light blue flag with a golden Moon in the center, represented by a yellow circle. The light blue field represents the sovereignty and the Moon peace. The full Moon is good for the harvest and planting crops and is believed to be holy.

The flag of Uzbekistan is also interesting. It has three horizontal bands. The top is light blue, the middle is white and the bottom one is green. The white band has two very thin bands on the top and bottom of it. The top band contains a white crescent Moon and 12 stars. The blue band is representative of water, the stars for the zodiac or 12 months, and a crescent Moon is the symbol of Islam but can also represent the birth of a new country.

The Moon is also associated with time and important events in religious calendars. For Muslims, the important religious festivals and events are determined in the Qur'an, but the actual date is usually determined by the Moon. Ramadan is a month of fasting during the daylight. This can also be known as the month of the Qur'an, as the holy book is recited from memory as much as possible during this time. It is also the time when the gates of Hell are shut and demons are powerless, making it easier to do good deeds. Good deeds are also believed to be special at this time, as Allah himself blessed this month.

Ramadan is usually calculated the way that the prophet Mohammed used. The thin crescent Moon, or hilal, is the start of the month, the following day being the beginning of Ramadan and fasting. This finishes when the hilal is seen again a month later and the festival Eid Al-Fitr begins. All the dates and special events in the Islamic calendar are based on Moon sightings, including the Islamic New Year and Eid-ul-Adha, or the Feast of Sacrifice, to remember the prophet Abraham's sacrifice of his first-born son, Ishmael, known as Isaac in the Bible.

In the Christian year, Christmas, commemorating the birth of God's son, is fixed, but Easter is not. Easter Sunday is the Sunday after the Paschal full Moon, although the Orthodox Easter usually occurs a week or two later.

This complex method was devised by astronomers and mathematicians of Pope Gregory XIII in 1583 following the replacement of the Julian calendar with the Gregorian one in October 1582. The point of the complex method of calculating the date of Easter was to try to replicate the season of the year and position of the lunar cycle on Easter Sunday as those occurring at the time of the resurrection of Jesus, believed to have been in A.D. 30. Easter Sunday generally falls between March 22 and April 25. Pentecost or Whit is determined by the Moon. It occurs on the seventh Sunday after Easter Sunday. Pentecost, or Whitsun, celebrated the Holy Spirit revealing itself to Jesus's followers.

The Moon is an important factor in the biodynamic agricultural calendar. This involves farmers using six different Moon rhythms every lunation (27–29 days). Biodynamic agriculture is about the Sun's effect on plant growth, with its influence on the tides. Liquid from the ground and soil rises to its surface at the time of the full Moon. More sap comes out of the plants when pruned at this time. The lunar period between the new Moon to the full Moon is the best time for pruning, sowing and inserting a shoot into a slit in a plant, from which it receives sap. Therefore, the period between the full to the new Moon is the optimum time to harvest, reap, plow and weed on the farm.

Although the Celtic year is based on the Sun, some festivals are celebrated at specific lunar times. Imbolc is all about life. This is when Persephone is believed to have returned from the Underworld and is around the end of January or beginning of February, at the time of the new Moon.

The festival of Beltaine celebrates fertility. This occurs towards the end of April to the beginning of May at the time of the full Moon. This is the time of year that includes May Day, maypole dancing and Walpurgis Night. Walpurgis Night occurs on the night of April 30, a time when winter ends and summer begins. It, along with Halloween, are the two nights when evil spirits are believed to be at their most powerful, and the veil between the spirit world and that of humans is supposed to be at its thinnest. On Walpurgis Night, witches gather in groups or covens to indulge in orgies and spell-casting. The most well-known of these gatherings is in Germany, at the top of the Brocken Mountains.

The festival of Lammas starts at the end of July or the beginning of August at the Harvest full Moon. It is also known as the Feast of the Sun God, Lugh. This can also be the time of the harvest and the waning of the Sun, although the harvest is properly celebrated later, in September, at the autumnal equinox.

Samhain is celebrated on October 31. It is also known as Halloween, All Hallow's Eve, All Soul's Night and the Feast of the Dead. Samhain usually occurs on a night when the Moon is dark. It symbolizes death and rebirth in the Celtic year. Like Beltaine, the veil between the spirit and human worlds is thinnest then, and it is a time when magic is especially potent, when people can try to talk to the dead and see into the future.

The gravitational pull of the Moon affects the ocean tides. The Moon's gravity creates a bulge of water that is being pulled away from Earth towards the Moon. For example, high tides occur when Earth spins through a massive tidal bulge at the equinoxes. The tides can symbolize give and take, mutual action, balance and opportunity. Some people believe that as the tide ebbs and flows, so does the soul.

King Canute, or Kund (995–1035), who by 1016 was king of England and by 1028 was king also of Denmark and Norway, was associated with the tides. Canute was a real king believed to have been pious and wise. He is supposed to have over-heard sycophantic subjects stating that he could actually tell the tides when to ebb and flow. He was only too aware that this was not possible, and this kind of power was God's alone. He decided to show his subjects that he did not have this power, and got some of them to bring his throne to the edge of the sea. He then sat upon it. He tried to tell the waves to stop coming closer when the tide was coming in. He knew full well that he did not have the power over the oceans and had intended to show his subjects that his power and that of other kings paled into insignificance when compared to the fearsome omnipotence of God.

This is likely to have just been a story, as there is no historical evidence to support this.

All nocturnal animals and birds are associated with the Moon simply because they are awake during the night, like the Moon itself. They are also associated with Satan, perhaps somewhat unfairly, as they are believed to be afraid of the light. The Chinese believed that foxes were the ghosts of the dead, probably because they sneak around at night. The badger is regarded as a yin creature (associated with the Moon) that, in Japan, is credited with having enchanted powers, while the bat can be regarded as the personification of Satan, being a fallen angel with its wings. South American vampire bats, which actually suck blood from mammals at night, have given other bats a bad name, associating them with Dracula and vampires, though most others feed on insects and fruit.

The ancient Egyptians believed that the humble mouse was originally shaped from the mud from the river Nile. They thought that the mouse's liver waxed and waned in tune with the lunar phases. As they are fearful creatures that seem to like the darkness, they are thought to have dark supernatural power. The wolf, as we saw, is also associated with the Moon and the werewolf. Hindus believe that the Asvins, who represent both night and day and light and the dark, resurrect the quail in springtime to stop them from being swallowed by the wolf in the winter, so the wolf swallows the more favorable, pleasant season of spring. It, like the fox, was believed in ancient times to be somewhat an eerie, spectral creature that scared people senseless. Tigers in southern China also had a fearsome reputation. People could be transformed into 'weretigers' similar to werewolves; however, devilish creatures were believed to be afraid of tigers, so they were not always regarded in a negative way.

For the Hindus, the antelope is a lunar creature. Chandra, the Moon goddess, rides in a chariot pulled by antelopes. The owl is also connected with ghosts, per-haps because they move so silently. The barn owl has a white face, which is often described as ghostly. In ancient China, the owl was associated with the purveyors of metal, reigning over the special days when they crafted enchanted mirrors. Cats, who go out at night to hunt, are associated with night and darkness, slyness and dark magic, as cats are often portrayed with witches, whereas in Japanese mythology, black cats could control women's bodies, and so are therefore inauspicious creatures and can be associated with spiteful djinns in Islam.

Because the horns of a ram are spiral shaped, they are linked to lunar goddesses. Because they contain parts that grow and shrink, spirals, and anything that is spiral in shape (like some shells) reflect the lunar waxing and waning. Some objects, such as beehives, are lunar purely because they are linked to the Moon goddess Artemis/Diana. Deer are also linked to this particular lunar deity as well as to Aphrodite and Athena, as are stags and hunting dogs. The frog and toad are both lunar creatures, the three-legged toad living in the Moon representing its three phases. The frog is supposed to bring rain, and being a wet creature, lunar, by reigning over the waters, like the Moon itself, and by being a protector of lunar gods and goddesses Fish (Fig. 6.4), like the serpents, have control over the waters, as does the oyster.

The crocodile in Egyptian myths cries when he swallows the Moon, thus the phrase 'crocodile tears.' The crab, however, is linked to the waning Moon. Pigs have no obvious lunar connection but were the sacrificial animal for Egyptian Moon festivities (Fig. 6.5), and the dragon, a solar creature, can also represent the potency of the waters.

The magpie has a tenuous connection to the Moon, as the Chinese version of the western Valentine's Day honors the reunion of two separated lovers for one single night of the year by a bridge created from magpies high up in the sky (Fig. 6.6). This only occurs once each year in the seventh month of the lunar year.

Even weddings have a lunar connection; they are symbolic of a spiritual union of the Sun and Moon, the solar bull and lunar cow, which, in turn represent fertility of the newly married couple, animals, crops and of the marriage of heaven and Earth (heavenly rain and earthly soil both being requirements for a successful harvest). Hot cross buns, being round in shape, eaten at Easter time can be symbolic

Fig. 6.4 The Moon's association with water also links it to fish. (Illustration by the author)

Fig. 6.5 Pig sacrifices were part of the Egyptian Moon festival. (Illustration by the author)

Fig. 6.6 In China, magpies are part of the mythology of the sky (and romance). (Illustration by the author)

of the full Moon with the cross representing its four quarters. Even a pack of ordinary playing cards has a lunar connection, as the 13 cards in each suit each represent a lunar month. The horseshoe, a traditional symbol of good luck is crescent-shaped and represents the Moon and lunar deities and the horns of the lunar cow,

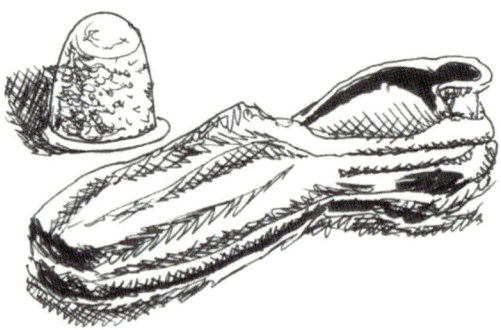

Fig. 6.7 Fairies were strongly associated with the Moon and belief in them continued even into modern times, for instance with the fairy shoe tradition. (Illustration by the author)

invoking her protection. The onion is believed to have the power to avert any bad luck, especially from destructive lunar powers.

Twilight and nighttime are obviously associated with the Moon. Twilight is a time of unease and transition—neither the light of the day and the Sun nor the dark of the night, the realm of the Moon. This uncertain state was believed to be a time for the Sidh when magical creatures such as fairies are supposed to appear. The silver bough or branch of an apple tree is believed to connect our human world and the realm of fairies, silver being associated with the Moon.

Although, by the early eighteenth century—a time of scientific discoveries and reason, with the English scientist Sir Isaac Newton's (1642–1727) laws— the educated ceased to believe in fairies, they did not die out in legends and stories. A fairy shoe measuring 2 in. (5 cm) in length was supposedly found in County Kerry, Ireland, in 1834. The tiny shoe displays actual signs of wear on the sole (Fig. 6.7).

Fairies were in fact taken seriously by Sir Arthur Conan-Doyle (1859–1930) in 1920. Conan-Doyle was a surgeon born in Edinburgh but was most famous for his Sherlock Homes stories about a fictional but extremely logical detective. Although a man of logic and science Conan-Doyle developed an interest in spiritualism. Around the time of the Great War (1914–19), spiritualism and séances were very popular, as people wanted to believe in some kind of afterlife and the possibility of communicating with the dead; fairies were part of this whole mindset, being part of another, more peaceful world where the possibilities were endless. A story involving Sir Arthur Conan-Doyle and fairies captured the imagination of the people in 1920. Two young girls in England took five photographs of fairies known as the Cottingley fairies in 1917. The photographs were used in a magazine article by Conan-Doyle and came to the attention of the public. The reaction was mixed. However, in the 1980s two women confessed that four of the photographs were faked. But one of the young photographers, Frances Griffith, at the end of her

Fig. 6.8 Nighttime and by extension the Moon is symbolized by a poppy. (Illustration by the author)

life stated that the fifth photograph was not fake and that she really did play with fairies as a child. By making these assertions, she kept alive a whole host of other possibilities.

Thunder can be connected with lunar changes, and obviously the Moon is associated with night, when it looks like the largest and brightest celestial object in the dark sky. The goddess of night, Nyx, is usually depicted in a dark, star-studded costume, sometimes holding two babies, the white one symbolic of sleep, the black one of death. Night can also be visually represented by an owl, black wings or poppies (Fig. 6.8).

Chapter 7

Mars, the Next Frontier

Mars is known as the Red Planet, because it is covered in reddish dust from iron oxide. This red color prompted the Romans to name the planet after their blood-lusting god of war.

Like the other classical planets, Mars was known to the ancients. Copernicus had mentioned the planet, and Galileo was the first person to observe it through a telescope. Mars is similar to our own planet, with volcanoes, polar caps, seasons and an iron core. Mars contains the largest known volcano in our entire Solar System, Olympus Mons. It has two known moons and takes 687 days to travel around the Sun. Mars's day length is also similar to ours, at just over 24 h. It is prone to storms that cover almost the entire planet with a jacket of reddish dust. Its gravity is around a third that of Earth's, and its orbit is more elliptical. It has lost practically all of its atmosphere (Fig. 7.1).

There has been much speculation about Martians and "little green men," perhaps wishful thinking on our part. Despite these optimistic thoughts, Mars has generally been regarded as a world devoid of life-sustainable conditions. On Christmas Day, 2003, the launch of the *Beagle 2* probe was televised. Unfortunately for everyone but especially the leader of this ESA project, the enthusiastic Dr. Colin Pillinger, the mission failed. Despite this, the whole event increased public awareness of space exploration.

However, there has been some relatively recent evidence suggesting there might have once been life on Mars. Spirit, NASA's Mars rover, found evidence of water, and over 10 years ago, NASA discovered frozen expanses of water similar to that in our Antarctic. Channels have been discovered that change according to the season, potential evidence of some sort of liquid, so perhaps there is life on Mars, after all?

© Springer Science+Business Media New York 2015

R. Alexander, *Myths, Symbols and Legends of Solar System Bodies*, The Patrick Moore Practical Astronomy Series, DOI 10.1007/978-1-4614-7067-0_7

Fig. 7.1 Mars with one of its polar caps. (Image courtesy of NASA)

Scientists have deduced that because bacteria can survive within permafrost (soil and ice frozen together) in Antarctica for millennia, then it is even possible that life could lie frozen on Mars, as some Martian terrain is similar. Mars is our next frontier, with an exciting joint NASA/ESA mission planned to go to the Red Planet in 2018. On August 6, 2012, NASA celebrated successfully landing a robot on Mars after taking 9 months to travel the 100-million-mile journey. The robot is called Curiosity and is on a 2-year mission to discover if Mars was once capable of supporting life and to see if there is any current microbial life from sampling and analyzing Martian rocks. This robot is paving the way for humans to visit Mars.

U. S. President Barak Obama is committed to a manned mission to Mars by 2030 and announced that another robot will be sent to Mars in 2016, to try and ascertain how all the planets were formed. In February 2013, a multi-millionaire named Dennis Tito announced a mission to Mars planned for 2018, around the same time as Inspiration Mars Foundation proposed that a married couple could go to Mars and back in 501 days in a tiny spaceship about the size of a caravan. New information is being drip-fed to the public via the news using data from the Curiosity rover. The data shows that life could have existed on Mars; as far back as four billion years ago, this planet had a thick atmosphere and was a warm, wet place with surface water. However, its magnetic field collapsed, and its atmosphere was destroyed. The Red Planet once may have had super volcanoes on its surface, as revealed by giant holes in its surface today. These volcanoes would have erupted and the volcanic ash blotted out the Sun to form the thick atmosphere.

Recently, it was revealed that over 200,000 people have applied for a one-way trip to Mars, which is intended to be turned into a reality television show. However, does this show public interest in new frontiers in space or their own personal desire for immortality?

In November 2013, India successfully launched its own mission to Mars, which will hopefully reach the Red Planet in September 2014. The probe will spend 9 months examining its geology and atmosphere. In the same month, NASA announced that a new spacecraft, *Maven*, will orbit Mars. It will take 10 months to reach its destination, and it will try to discover why Mars turned cold and dry after being warm and wet. It will also try to find out how this planet lost its rich atmosphere. It is currently thought that the solar wind ripped most of it away. *Maven* will orbit Mars to study its atmosphere, which is believed to be very weak. The Space Race appears to be continuing, with the political will for exciting future exploration of Mars, but the biggest obstacle is financial. Perhaps emerging economies such as India and the private sector will take up the gauntlet.

Ares/Mars was the god of battle and war. He loved battles and took great pleasure in seeing men being killed, which made him extremely unpopular among the other deities, with the exceptions being his sister, Eris or Strife, Aphrodite and Hades. In one myth, his parents are supposed to be Zeus and Hera, while in another, Flora/Chloris, the goddess of spring, was supposed to have provided Hera with an enchanted plant that enabled her to conceive without the help of her husband. The result of this floral conception was Mars, ironically supposed to be a symbol of masculinity.

The Greek Ares enjoyed inciting violence. He drove a chariot led by four fierce horses, who exhaled fire from their nostrils, rather like dragons. Despite being god of war, he was not as brave as Athena/Minerva, his half-sister, who was the favorite child of Zeus. During the Trojan War, he fought on the side of the Trojans, led by the brave Hector, brother of Paris, against the Greeks led by Diomedes, the side Athena was on. Athena managed to deflect the spear of Diomedes so that Ares himself became wounded, and when the gargantuan and terrifying monster Typhon appeared, the mere sight of him was so fearsome that all the gods took flight to Egypt and transformed themselves into animals. Ares metamorphosed into a boar. Athena alone stood her ground, proving her bravery and putting all the others to shame and forcing Zeus to come out and fight.

Mars was not alone on the battlefield. He was usually accompanied by Eris/Strife and his two sons, Deimos (Fear) and Phobos (Panic).

Fig. 7.2 Ares or Mars transformed into a boar to slaughter his rival Adonis. (Illustration by the author)

Mars was unmarried in most myths but loved Aphrodite, the mother of his children Deimos, Phobos, Eros and Harmonia. They had a tempestuous affair, which ended when Hephaestus, Aphrodite's husband, ensnared them in a fine bronze net when having sex, making him the object of envy of the gods Hermes and Poseidon but a laughing stock by the other deities. He still loved the beautiful Aphrodite, but she got over him almost immediately, falling in love with the handsome young Adonis. Aphrodite caught Ares and Eos having sex and was furious. She was angry and jealous and chose to curse Eos with a relentless sexual desire for beautiful young men. Ares revealed his dark, jealous, murderous side by transforming again into a boar and killing the unfortunate Adonis (Fig. 7.2). The Roman god Mars was married to Bellona, their goddess of war. Sometimes, she is presented as the sister of Mars, though.

When the 24 imposing giants rebelled and attacked Mount Olympus in retaliation for Zeus's overthrow of the Titan dynasty and imprisonment of some of their siblings in Tartarus, a bloody battle commenced. Despite being in his elements, Ares fought against the giant Ephialtes and found himself in a perilous situation until Apollo rescued him by shooting the unlucky giant in one of his eyes and killing him. However, Ares showed considerably more courage and valor after this timely rescue by indulging in his most pleasurable pastime of murdering the remaining giants by stabbing them with his famous spear alongside Zeus and his deadly thunderbolts, although Heracles delivered the final fatal blows.

Ares did admire the mortal, Paris. Although the son of Priam, who was ruler of Troy, he was estranged and had a humble upbringing. However, Paris was incredibly handsome and clever. He enjoyed setting up bullfights with the cattle of his foster-father, Agelaus. He discovered a champion bull that kept on defeating the others, so he organized fights with this champion bull and those of other herdsmen. His bull continued to outshine the others, so he challenged anyone to defeat his bull, offering a crown of gold for the winning animal. Ares found it highly amusing to transform himself into a bull and defeated Paris's champion. Paris immediately placed the golden crown on Ares' horns. This act made Paris a favorite among the gods, leading to him being given the somewhat dubious honor of judging who was the most beautiful goddess.

The Roman god Mars evolved from a previous agricultural and storm deity, who slowly turned into the violent, battle-loving god that was more similar to the Greek Ares, reflecting Rome's own expansionism. In one version of the mythology, he was married to the goddess Nerio, with whom he sired Romulus and Remus. However, there is a much more controversial version of this myth that highlights the double standards of the gods.

Rhea Silvia was the daughter of the overthrown King Numitor of Alba Longa. She was a former princess turned Vestal Virgin, so that she could not bear children who would be potential heirs and threats to her uncle's recently stolen throne. Rhea Silvia was, of course, also beautiful, which combined with her virginity made her irresistible to feisty Mars. Rhea Silvia left Vesta's temple to collect water. She fell asleep by the river in Mars's hallowed wood. Mars saw the innocent, defenseless virgin sleeping, lusted for her and raped her. Rhea Silvia continued to sleep throughout this ordeal, dreaming that she was in the city of Troy, and dropped her hair pin, whereupon two trees grew. One tree grew large enough to shelter the whole world, a metaphor for the Roman Empire. Eventually, Rhea Silvia discovered that she was pregnant and gave birth to twin boys, Romulus and Remus.

The twin baby boys were placed in a basket on the river Tiber and were expected to perish. However, this being mythology, nothing was that simple. The basket was washed up on the riverbank near a cave called the Lupercal, meaning "Place of the Wolf." A she-wolf discovered the helpless babies and sustained them by giving them her milk. They were eventually found by a shepherd, Faustulus, who brought them up. Romulus and Remus, as adults, set up their own city, Rome, the center of the future Roman Empire. The brothers quarreled, which led to a fatal fight that left Remus dead. Because Romulus's father was the god Mars, the city was granted special status. Romulus set up defenses at Rome's Palatine Hill, sacrificed to the gods and gave people laws. However, he discovered that there were not enough women of child-bearing age to ensure the survival of the new state of Rome. So he staged a harvest festival, which attracted people from nearby, including many young and beautiful Sabine women. Towards the end of the festivities, the young Roman men captured and sexually attacked the Sabine women. This became known as the Rape of the Sabine Women, which led to war.

Ares, being fiery, tempestuous and impulsive, was furious that Poseidon's son, Halirrhotius, had tried to rape his daughter, Alcippe. Impulsively, he murdered Halirrhotius, who he forgot was not a mere mortal; he had murdered the son of a fellow Olympian deity. Poseidon was, understandably, furious in turn. He held a jury of the gods to judge whether or not Ares was guilty of murder. Despite his unpopularity among the other deities, Ares managed to persuade them that Halirrhotius deserved his fate. The people of Athens then named the location of this first-ever jury trial Areopagus, meaning "Hill of Ares."

Mars, according to legend, fell in love with Minerva, the goddess of wisdom. In Rome, Minerva was depicted as battle-loving, wearing armor and a helmet and carrying a spear. Mars desired her, but she was a virgin goddess and refused his

Fig. 7.3 Woodpeckers are associated with Mars. (Illustration by the author)

advances. He decided to get help from the goddess of the New Year, Anna Perenna, so that he could marry Minerva. Anna Perenna, however, decided to trick the amorous Mars and pretended to be Minerva at their wedding by wearing her clothes and a veil that covered her face. He was tricked into marrying Anna Perenna, who was an elderly goddess and not nearly as desirable. This legend was kept alive by song. During Anna Perenna's annual festival on March 15, songs were sung, praising her famous deception.

The weekday Tuesday is associated with Mars, the French word for the day being *Mardi*. Mars is personified as an aggressive man holding a deadly spear. He is associated with the metal iron, the color red and the wolf, as a female wolf suckled his children, Romulus and Remus. Another of his symbols is the woodpecker, because of the bird's perceived anger and destructive activities (Fig. 7.3).

The Mars candy bar and company were, in fact, named after the Mars family rather than the Roman god of war. However the company has capitalized on its planetary connection by its choice of other confectionary product names, including Planets, Galaxy, Milky Way and Magic Stars. The popular, chocolate-coated sweets or candy known as M & M's were the choice of the first U. S. astronauts on the space shuttle, in 1982, thus cementing this cosmological connection.

There are countless Martian craters, but only around 1,000 have been given a name. Most of them are named after places on Earth. A few have space connections, like Canaveral after Cape Canaveral, the home the Kennedy Space Center and the Cape Canaveral Air Force Station from which many famous spacecraft were launched, such as the first U. S. Earth satellite in 1958 and the first spacecraft to orbit Saturn in 2004. The Goldstone Crater is named after the Goldstone Observatory in California, which tracks radar observations from Venus and which, back in 1961, revealed the planet's slow rotation.

The Texas Crater is named after the U. S. state, the home of NASA's Johnson Space Center in Houston, which contains a massive vacuum chamber, the only one cold enough and big enough to form an environment similar to that of space itself for the James Webb Telescope, the intended successor to the Hubble Space Telescope. (The James Webb Telescope will be sent to an orbit about a million miles from Earth possibly as soon as 2018.) The Jodrell Crater is named after the Jodrell Bank Observatory in England, home to the Lovell radio telescope. The observatory was established back in 1945 by the recently deceased Sir Bernard Lovell.

The Eagle Crater is named after the lunar module Eagle from the *Apollo 11* mission that first landed men—Neil Armstrong and Buzz Aldrin—on the Moon. Finally, the Mariner Crater is named after *Mariner 4,* the fourth in a series of flyby spacecraft for the exploration of planets. This one appropriately reached Mars in 1965 and took the first close-up images of the Red Planet.

Many are named after the 'usual suspects,' those who already have a lunar crater named after them. These include Columbus, Galileo, Lockyer, Kepler, Herschel, Copernicus and Newton. (See the chapter on the Moon for more details). Some are named after writers of fiction, which include the Burroughs (1875–1950), the creator of the character, Tarzan and the Mars adventurer John Carter. The Wells Crater is named after the English science fiction writer H. G. Wells (1866–1946), who wrote, among other novels, the *War of the Worlds* and the *First Men in the Moon.* The Roddenberry Crater is named after Gene Roddenberry (1921–91), the American television screenwriter responsible for bringing fictional space travel, in the form of the *Star Trek* and *Star Trek: The Next Generation* television series, to our living rooms. The Sagan Crater falls into this category of writers, named after the American writer, astronomer and astrophysicist who narrated a television series in 1980 and began the search for and study of life beyond Earth.

Other craters are named after explorers, adventurers and even the vessels in which they traveled. The Beagle Crater is named after the ship the HMS *Beagle,* in which Charles Darwin journeyed. His geological discoveries made during the voyage of the *Beagle* eventually lead to his theory of evolution. There is also a Darwin Crater named after the HMS *Endurance,* the vessel that was made famous by the British Antarctic explorer Sir Ernest Shackleton, although it became trapped by ice that eventually crushed it. (Shackleton does not have a Martian crater named after him, but he does have a lunar crater in his name on the far side of the Moon.) Other craters in this category include Columbus, Magellan and Ejriksson, the latter named after the Icelandic explorer Leif Ericson (356–323), who probably discovered North America before Columbus.

Most of the other named craters honor some lesser known astronomers and scientists, including William Hammand Wright (1871–1959), the American astronomer who made images of Mars in different wavelengths from which he calculated that the planet has a very thick atmosphere of around 60 miles (100 km) deep. The Stokes Crater is named after the Irish Sir George Gabriel Stokes (1819–1903), the mathematician and physicist.

The Lowell Crater is named after the American astronomer Percival Lowell, who believed that there was another planet beyond Neptune and conjectured that the canals on Mars were vegetation fed by water from the polar ice caps—an interesting theory but refuted by the imagery from *Mariner 4* in 1965. Giovanni Schiaparelli also has a Martian crater named after him (1835–1910). He was an Italian astronomer who claimed to see 'canals' on Mars, which started much speculation about possible life on the Red Planet. The Huggins Crater was named after Sir William Huggins (1824–1910), an English astronomer who used spectrum analysis and revealed that stars are made from the same elements as Earth itself and our Sun.

One crater is called Nereus. In Greek mythology Nereus was the old man of the sea, a shape-shifter. He was the father of the 50 Nereids. Another crater is Heimdall, named after the mythological watchman for the Norse gods. He saw and heard everything and guarded Bifrost Bridge, which connected the heavens with Earth.

Only craters larger than 60 km (36 miles) that are interesting in terms of further research are named after famous scientists, astronomers, explorers and writers. Craters that are of scientific interest but are smaller than this are named after places, ships and minor mythological characters. There are many, many more named Martian craters than are featured in this book.

Mars has two known moons, Deimos and Phobos. These were both discovered by the American astronomer Asaph Hall in 1877. Both moons are small and irregularly shaped. Phobos, the larger moon, is less than 27 km (17 miles) across, while Deimos is a mere 15 km (9.3 miles) across and is further away than Phobos. It is conjectured that both moons could be rogue asteroids, captured by Mars's gravity (Figs. 7.4 and 7.5).

Asaph Hall chose the names for the moons that he discovered, which he believed were after horses that pulled Mars's chariot. However, Homer gives the names of Deimos and Phobos, Fear and Panic, to Ares' sons. The word phobia comes from *phobos,* or "panic."

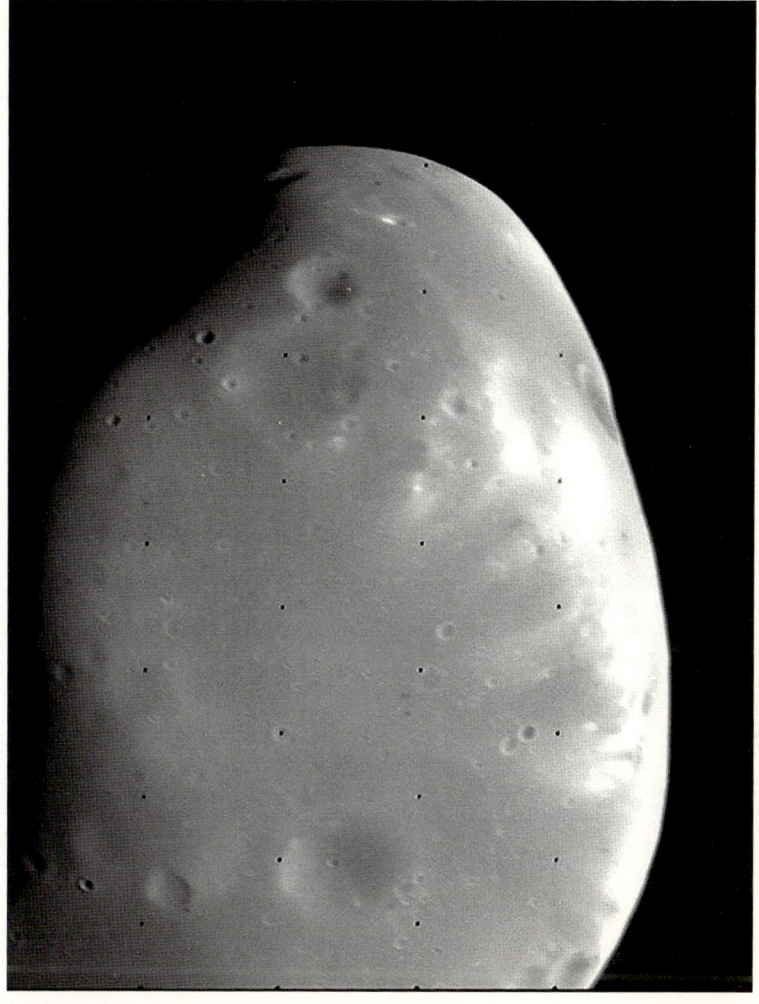

Fig. 7.4 Martian moon Phobos. (Image courtesy of NASA)

Fig. 7.5 Martian satellite Deimos, the smaller of the two. (Image courtesy of NASA)

Chapter 8

Doom and Gloom: Comets, Meteorites, and Asteroids

The word comet is derived from the Greek *kometes,* meaning "long-haired star." Comets are bodies comprised of ice and dust and travel around the Sun in elliptical paths. They have been nicknamed dirty snowballs. They are visible from Earth when they get close to the Sun. The Sun's extreme heat turns their ice into gas and creates their twin tails, one of dust, the other of gas, which always point away from the Sun. Meteor showers occur when Earth travels through the scattered fragments of comets left behind as they travel around the Sun.

It is believed that comets at the edge of our Solar System do not have a tail. When comets pass the planet Jupiter, they produce their twin tails and become illuminated. Comets can be either short- or long-period types. Halley's Comet is an example of a short-period comet. Long-period comets can travel in their orbits anywhere from 200 to a million years. They are on the very edge of the Solar System, where they can barely feel the Sun's gravity.

Comets meet their death in different ways. Some will reach their end by a violent impact as they approach the Sun, while others will merely disintegrate into dust.

No one particular person is credited with their discovery. Astronomers in Babylon and China understood the movements of some of the planets as long ago as 800 B.C., when Homer wrote his *Iliad* and *Odyssey.* However, it was centuries later, around 200 B.C., that astronomers in China noticed, amazingly, that the tail of a comet always pointed away from the Sun regardless of the direction of travel of the comet itself. By the second century B.C. Chinese astronomers were able to categorize comets into almost 30 different types. In November 1680, the famous English astronomer Sir Isaac Newton saw a comet disappear behind the Sun. This made him think that there must be an invisible force working that he named gravity. Newton and Sir Edmund Halley were also able to predict when comets would

© Springer Science+Business Media New York 2015   133
R. Alexander, *Myths, Symbols and Legends of Solar System Bodies,* The Patrick
Moore Practical Astronomy Series, DOI 10.1007/978-1-4614-7067-0_8

Fig. 8.1 Christmas Star. (Illustration by the author)

appear. Halley, also an English astronomer, discovered that comets traveled in an elliptical path around the Sun. He also discovered that one particular comet had a 76-year orbit. This comet is named after him, Halley's Comet.

Comets throughout history have been regarded as both auspicious and inauspicious omens. The Christmas Star (Fig. 8.1), which the Magi followed to Bethlehem to find the baby they believed to be the son of God, was regarded as an unquestionably auspicious sign. The warrior Genghis Khan is supposed to have regarded comets as a message from the heavens for him to declare war and fulfill the preordained predictions of his life. Some peoples believed that comets were symbolic of the presence of the gods up in the heavens. Some Muslims believed that comets were lumps of burning wood hurled by benevolent angels at malevolent spirits when they lingered too near the famous gates of heaven. Other people believed that if a wish is made when a comet falls, then it will be granted, almost as if the comet is a god or has the power itself to make wishes come true.

There is even speculation that the British hero, King Arthur was the personification of a comet, as his battles occurred during meteor showers and his sword was supposed to have been especially shiny. Comets have sometimes been believed to have been used by wizards or single-eyed giants to make swords, as ancient peoples of different cultures used mythology to try to make some kind of sense out of something inexplicable.

As comets appear suddenly out of nowhere, they threaten the perception of an ordered cosmos. The discoveries of both Sir Isaac Newton and Sir Edmund Halley have demystified comets or at least their sudden appearance, but comets will always provide material for stories and mythology. For the English King William I, also known as William the Bastard or, more famously, William the Conqueror, Halley's Comet was such an auspicious sign that it became immortalized in the Bayeux tapestry, embroidery on an epic scale depicting the Norman victory led by William over the English King Harold's army in 1066. Comets were once known as Sky Wolves or Sky Foxes and meteors as meteor dragons, perhaps because of their tails, suggesting fear, wonder and awe.

Comets have been linked to superstitions concerning the end of the world, doom, death, destruction and downright bad luck. Some linked the appearance of a comet with the presence of the devil or a malignant influence. The Persians believed that the Peri, a beautiful and malevolent sprite, was responsible for comets and eclipses that caused the failure of crops. In 1577 in the Ottoman Empire, an astronomer known as Takiodene observed a comet. He told Sultan Murat III that it predicted Ottoman victory, which was likely to have been what he wanted to hear; this victory did not happen, so the sultan closed down the observatory.

In 1159 B.C. some disastrous weather event occurred throughout the world, according to dendrochronological (tree-ring growth) evidence. There was an over-abundance of rain, with 18 consecutive summers without Sun, just endless rain. It is now conjectured that this was caused by a comet traveling across the sky that enrobed Earth in a coating of dust. This is likely to have had a significant effect on different places on Earth, leading to a change of dynasty in China, the breakdown of Greek civilization, and the evacuation of inhabitants of Dartmoor, England, within a few generations.

Halley's Comet was not exactly lucky for Harold Godwinson, the richest man in England. Harold claimed that King Edward the Confessor had promised him the English throne before he died. Harold had himself crowned king of England on the same day as Edward's but was defeated in the same year by William the Conqueror at the Battle of Hastings in 1066. Harold was shot in the eye with an arrow, according to the Bayeux tapestry. Halley's Comet appeared in the sky soon after William's victory and was then associated with change, earning it the name 'The Terror of the Kings.'

Some believe that the ninth Aztec king, Montezuma, saw a comet that he thought was symbolic of the end of his kingdom, so he did not fight the invading forces led by Hernando Cortez in 1519. Montezuma must have thought that his destiny had been written in the heavens.

The Great Comet of 1811, which was discovered by the French amateur astrono-mer Honore Flaugergues (1755–1835) was conveniently blamed for Napoleon's defeat at the Battle of Waterloo in 1815. This comet was observed by William Herschel and would have been visible, even to the naked eye, for around 9 months.

More recent was the comet Hale-Bopp, discovered by two American astrono-mers, Alan Hale and Thomas Bopp. This comet was so bright that it was visible to the naked eye for a whole year. It was visible to more people than any other comet so far. It was enormous, with a huge, icy core of 25 miles (40 km) in width, with twin tails that expanded to a vast 62 million miles (100 million km) in length. This comet might be remembered for the mass suicide of members of the American cult called Heaven's Gate. It all started with a photograph of the comet that included another space object vaguely resembling Saturn. Some, like the Heaven's Gate members, chose to interpret the object as an alien spaceship. This fitted in with their belief that they were to be taken by aliens in the spaceship that followed the comet. Thirty-nine members killed themselves to speed up the journey of their souls to an alien world. (The Saturn-like object was believed to be a background star, accord-ing to astronomers.)

Comet Ison C2012S1, nicknamed 'Kerfuffle,' was predicted to be the brightest comet ever to be recorded in human history. It was purported to be brighter than the Moon and would be seen in the daytime. This turned out to be a spectacular celes-tial disappointment for many astronomers, who were eagerly awaiting the 'comet of the century.' A very special comet at 4.6 billion years old, Ison came from the Oort Cloud. It was formed at the beginning of our Solar System in the same way as the planets. Therefore, if we could discover how this comet was formed, we could work out how the planets in our Solar System were formed.

However, Ison apparently did not survive its close up and personal grazing of the Sun's corona and is now believed to be dead. On a more positive note, a vast amount of invaluable comet data was gathered from this event, so hopefully Ison did not die in vain.

In November 2014, the European Space Agency's Rosetta spacecraft should reach the Comet 67P/ Churyumov-Gerasimenko and will try to land a probe on its surface. Hopefully this will help us learn more about the comet's solid nucleus.

Comets also have their place in mythology and stories. In the Book of Genesis in the Bible, the destruction of Sodom and Gomorrah, described as rain on burning sulfur, has been blamed on a comet. Then there is the Greek myth of Phaeton, a son of Apollo, who destroyed part of Earth. The unlucky Phaeton 'borrowed' his father's chariot in which Apollo, god of light and the Sun, crossed the sky every night. Phaeton lost control and fell out of the chariot, plunging to his death and burning part of our own planet.

In Greek mythology Electra, supposedly one of the Pleiades, mother of Dardanus, the ancestor of the Trojans, is believed to have vanished just before the Trojan War, as she could not bear to witness the destruction of Troy. She is said to have showed herself but was disguised as a comet.

Comets contain water and complex chemicals. It has been discovered that the craters on the Moon contain ice from frozen water from comets. If even a small piece of a comet collided with Earth, it could destroy a whole area and could cause problems for all of our planet by coating it with dust, blocking out the Sun for a time.

There is also much superstition connected to meteorites, rocks that have fallen from space itself. It is not surprising that the ancients believed that comets or meteorites were regarded as messages from a higher power or deity, as they literally fell out of the sky. In Yorkshire, England, people once believed that shooting stars, which meteorites can appear to be from, were the personification of souls coming to Earth to bring new babies into the world. In the Islamic faith, the black-colored stone at the shrine of the Ka' aba is believed by some to have come from the sky and is a meteorite.

Meteorites, celestial objects that hit our atmosphere and usually burn up, have great destructive potential. The Tunguska event of 1908 was the largest recent impact event on Earth. It produced a massive fireball and completely flattened about 770 square miles of forest. It also lit up the sky as far away as London. It is still not clear if this was a comet, a meteorite, or something else.

On February 15, 2013, a meteorite hit Russia's central Ural Mountains. It caused a sonic boom that, in turn, caused glass from buildings to break and over 500 people to be injured. Factories and buildings were damaged in six cities. This meteorite was unexpected, too small to be spotted in advance, revealing how precarious our lives are and how vulnerable our planet is. Some people said that it was like a bomb going off. This has become known as the Chelyabinsk meteorite, after the Siberian city where it landed. It is believed that this meteorite was either deflected towards Earth by Jupiter's gravity, by collision or due to the Yarkovsky effect, where a slight force from the Sun, acting on photons, over time, can push an asteroid away from its usual orbit. This particular asteroid came out of the blue, but another asteroid, DA14, came close to Earth on the same day; scientists had been tracking this one for a year.

A fragment of the Chelyabinsk meteorite was retrieved from a lake in Russia's Ural Mountains. It is a small but very dense rock with a blackened surface and rounded edges. This is the first time that a meteorite has been filmed and then found. Gold medal winners at the 2014 Sochi Winter Olympics received medals containing tiny pieces of the Chelyabinsk meteorite, making them unique and special.

The Aztecs had a minor star god called Tzitzimime, whom it is conjectured is named after meteorites, as *Tzitzimime* means "large creature" or "thing falling from above."

It is believed that Earth was attacked by a huge number of comets during the period known as the Late Heavy Bombardment. These impacts are still occurring, but we are now being hit by only minute dust particles that contain water and carbon. Comets might have provided Earth's oceans and surface water, but if they gave us water, they could give it to other planets.

Comets are a reminder of both our significance—and insignificance—in our Solar System. We are reminded of our precarious existence but simultaneously of how lucky we are to live on a planet where conditions for our survival are just perfect—for now, at least.

The Asteroid Belt is situated between the planets Mars and Jupiter. The rocks and debris that are there are believed to be the pieces of a planet that had been unable to coalesce due to the sheer power and magnitude of Jupiter's gravity.

An asteroid is a small rocky celestial object or body that travels around the Sun. Most are found in the Asteroid Belt, although there is a residue of planetesimals in the Kuiper Belt, beyond Neptune.

Asteroids can be incredibly hazardous to Earth. It is believed that 65 million years ago, an asteroid that could have been as small as between 6 and 9 miles (10 and 15 km) hit our own planet. This impact annihilated almost half of our flora and fauna, leading to the demise of the dinosaurs. It would have had disastrous consequences, with fireballs setting fire to vegetation across the whole planet, extremely fast winds, storms, tsunamis and dust clouds, causing winter to last for months or even years.

More recently, in 1908 in Tunguska, what might have been an asteroid hit Earth and flattened a large area of around 3,300 square miles about (2,000 sq. km). Even today, some near-Earth objects such as asteroids remain perilously close to our planet.

Asteroids are too small, usually less than 600 miles (1,000 km), to be regarded as planets. The first asteroid to be discovered was Ceres, in 1801. In mythology, Ceres (also known as Demeter), was the daughter of Saturn and was the goddess of grain, crops and vegetation. She was associated with death because her daughter Proserpina/Persephone lived for part of the year in the Underworld, or realm of the dead, with her husband, Pluto/Hades. Ceres is the largest known asteroid at 540 miles (900 km) across, and it has been recently re-classified as a dwarf planet. Ceres is an asteroid dominated by ice.

In Roman times there was a festival called the Cerealia, which occurred in April for 8 days in honor of the goddess Ceres. On the final day of the Cerealia festival, the path of the Sun is symbolized by the release of foxes with torches tied to their tails (Fig. 8.2). Our word cereal originates from the name of the goddess, Ceres.

Fig. 8.2 Flaming torches represented the Sun in the Roman Cerealia festival. (Illustration by the author)

Some asteroids or planetoids are in the Kuiper Belt, beyond Neptune, such as Sedna and the recently downgraded planet, Pluto.

The second largest asteroid is Vesta at 300 miles (500 km) in length. In Greek mythology Hestia was the daughter of Cronos and Rhea and the sister of Zeus. She was the virgin goddess of the hearth and home; her Roman equivalent was Vesta. She had a temple dedicated to her in Rome, where a flame burned continuously, tended to by the Vestal Virgins. Vesta is the brightest asteroid, the only one visible with the naked eye. Around Vesta in the Asteroid Belt are its fragments, some of which have arrived on Earth and are composed of basalt.

We will hopefully learn more about the asteroids Ceres and Vesta soon, as the spacecraft *Dawn* is headed toward them on a mission.

The next two largest asteroids are Pallas and Hygiea. Pallas was a childhood playmate of the goddess Athena. While playing fighting games with her real weapons, Athena accidently fatally wounded the unlucky Pallas. Athena was devastated and called herself Pallas Athena as a tribute to her dead friend. Hygiea was the daughter of Asclepius, who apparently learned how to heal from the god Apollo. Asclepius is famous for curing ill people and was even able to raise the dead. However, he had to use blood from the dead gorgon Medusa in order to perform this task.

Sedna is another asteroid, discovered in 2003. It was the furthest object from the Sun to have been observed. It is a red celestial body, almost as red as the planet Mars, and it could possibly have its own moon.

Sedna was a character from Inuit mythology. There are different variations of this myth. In one, Sedna was a beautiful young woman who was married to a man

Fig. 8.3 The Inuit legend of Sedna imagined that seals formed from her injured fingers after she drowned. (Illustration by the author)

that went out hunting. This man turned out to be a raven. Sedna became distraught on discovering this and was inconsolable until her father agreed to collect her in his kayak to take her back home with him. The raven, understandably, was not happy about this and pursued Sedna and her father as they tried to escape. Sedna fell out of the kayak. Her father cruelly sliced off poor Sedna's fingers to save himself as his daughter tried to cling onto his kayak. She fell down, down to the bottom of the ocean, and her bloody fingers were transformed into whales and seals (Fig. 8.3).

There are hundreds of millions of asteroids in the boundary between the inner and outer Solar System. One of these asteroids is Eros. In mythology, Eros was the god of sexual desire, with Gaia and Tartarus for parents. Eros is capable of driving men and gods out of their minds with desire! The asteroid Eros is small, a mere 19 miles in length (31 km).

Asteroids are believed to contain within them the story of the beginning of our Solar System, although we have not yet been able to study them in detail. However meteorites are believed to be tiny bits of asteroids that have fallen to Earth and can be studied. There are different types of meteorites, and therefore different types of asteroids. Some meteorites contain valuable, pure metals, such as iron and nickel. Back in the heyday of Moon exploration, there was talk of mining asteroids for the precious metals they contain, when there is more money available for such potential missions. This concept has recently resurfaced, but with the idea of mining smaller 'mini moons,' or small asteroids that can be captured by Earth's gravity.

Fairly recently, asteroids have been discovered that have tails, like comets, but they are not comets. This has exciting implications for our own planet, our Solar System and beyond. Our Earth, with its large quantity of liquid water, was believed to have obtained this from comets, but now it seems it could have acquired water from asteroids as well. Some meteorites that have been analyzed were found to contain amino acids and the building blocks of DNA. They could have brought to Earth not only water but microbial life itself or the right chemicals for it. It is also theoretically possible that life could have formed on an icy planet from our outer Solar System.

Asteroids are potential symbols of creation and destruction. They could have physically brought life to Earth, but could annihilate that life if our beautiful planet was hit by a very large asteroid. President Barak Obama has challenged NASA to send astronauts to an asteroid by 2025, so we can learn more about them. Thus they remain on the political agenda.

Chapter 9

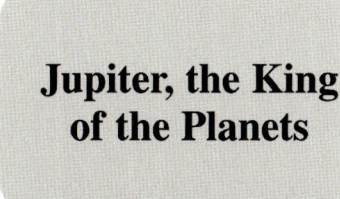

Jupiter, the King of the Planets

The Planet Jupiter

The planet Jupiter is much more than a mere gas giant. It is a gargantuan typhon of a planet, more than twice the volume and mass of all the other seven planets in our Solar System combined (Fig. 9.1). It has a diameter of 142,964 km (88,650 miles) compared to that of Earth at a mere 12,728 km (7,909 miles). Its Great Red Spot (GRS) is so large it could swallow up Earth and has become the most well-known storm in our Solar System. Robert Hooke noticed the GRS in 1664 but didn't know what it was. The GRS is a storm, or cyclone, and has been sighted on and off probably since the invention of the telescope.

Jupiter consists mainly of hydrogen and helium, and its clouds conceal a world that is anything but solid, so there is no whole or stable surface for a probe to land upon—yet another challenge for space scientists. It is believed that all the gas giants have a core of dust.

Jupiter has often been called a failed star because of its sheer size. Stars are also made up of hydrogen. Jupiter is large but not quite large enough to trigger nuclear reactions that, in turn, would cause it to produce its own light, like a star. Jupiter produces a lot of internal heat instead of merely reflecting back the heat of the Sun. It is conjectured that Jupiter's magnetic field is powered by metallic liquid nitrogen in its interior. This magnetic field is around 10,000 times stronger than that of Earth's. This field is so strong that it can pull at the Sun as it revolves around it and is believed to have caused the Asteroid Belt, made up of an unformed planet, as well as the numerous volcanoes on its moon, Io. Jupiter also has a system of rings

© Springer Science+Business Media New York 2015
R. Alexander, *Myths, Symbols and Legends of Solar System Bodies*, The Patrick Moore Practical Astronomy Series, DOI 10.1007/978-1-4614-7067-0_9

Fig. 9.1 The gas giant Jupiter. (Image courtesy of NASA)

surrounding it, discovered by the Voyager probes in 1979. Jupiter's rings are rather like flour—fine crystalline ice-water particles.

Jupiter is sometimes regarded as Earth's protector, as its powerful gravitational pull attracts many comets that might have traveled towards our planet and caused collisions.

Jupiter is said to rule the late-night skies, outshone only by Venus. It shines brightly, looking white to the naked eye. It has earned the nickname amateur planet, due to how easy it is to spot. It was known to the ancients as one of the five wandering stars.

A U. S. spacecraft departed in August 2011 for Jupiter and will take 5 years to get there. This craft is appropriately named *Juno,* after the mythological Jupiter's wife. It seems appropriate that the Romans named their largest planet after their supreme mythological deity, Jupiter.

It is generally accepted that the Italian astronomer Galileo discovered Jupiter's four largest moons—Io, Europa, Ganymede and Callisto—collectively known as the Galilean moons. He noticed and drew their progressions around their parent planet as early as 1610. The following year, the German astronomer Simon Marius claimed that he had discovered these four moons a couple of months before Galileo did. A dispute ensued. Galileo announced that Marius imitated his work and changed the date so that it created the illusion that he had found the four moons first. However, Marius was responsible for deciding to call the moons after the mythological lovers of Zeus/Jupiter, although it took around 350 years for these names to be made official.

Zeus/Jupiter, the God

Zeus was the youngest son of the Titan gods Cronos and Rhea. Cronus was afraid that a prophesy pronounced by his own brutally castrated dying father Uranus would come true. The wounded Uranus, with blood all around him, predicted that one of Cronus's own sons would overthrow him. This thought terrified Cronus. He was probably afraid that he could end up castrated, too. His solution was to swallow all of his children, even the girls, as soon as they were born, just to make sure the prophesy couldn't come true. Zeus escaped being swallowed thanks to his mother, Hera, who wrapped up a stone in swaddling clothes to resemble a baby and gave this to Cronus to swallow instead.

Zeus grew up on the island of Crete and was nursed by the goat-nymph Amaltheia. Eventually, Cronus discovered that Zeus was very much alive and came looking for him. Zeus transformed himself into a snake to remain hidden. When he was an adult, he found the Titaness goddess Metis, who provided a potion that caused Cronus to vomit up all of the children that he had swallowed. The Titans, led by Atlas, and the Olympian gods, led by Zeus, fought for a decade. The Olympians triumphed as Zeus released the those previously imprisoned from Tartarus, the 100-handed creatures and the Cyclopes, the latter of whom presented Zeus with his famous weapon, the thunderbolt. This legend supports the tale that Hades got an invisibility helmet and Poseidon his trident from the Cyclopes. Eventually, Zeus hit Cronus with a thunderbolt. 'Atlas was' NOT The only Titans that were left after lengthy battle were Tartarus and Atlas with the unenviable task of eternally holding up the heavens on his shoulders, although in a variation of this myth he holds up Earth.

Zeus became supreme leader of the Olympian gods. His realm was the sky, with Hades getting to rule the Underworld and Poseidon the sea.

Zeus had to fight additional battles to retain his power. In retaliation for imprisoning their siblings, the Titans, comprised of 24 furious and terrifying giants, attacked the Olympian gods. The gods fought off the giants, Zeus with his famous thunderbolts. The gods eventually won.

After this, Zeus had to face an even bigger challenge, literally in the shape of the gargantuan, fear-inspiring monster Typhon, the largest malformed fantastical creature that had been born, with snakes for hands and lower legs. When he approached Olympus, the Olympian deities ran away to Egypt, including Zeus, who transformed himself into a ram. Athena, goddess of wisdom, accused Zeus of lacking courage. Zeus then started to attack Typhon with a thunderbolt, wounding it. Typhon then removed Zeus's tendons from his hands and feet, severely limiting him so he couldn't release his thunderbolts. Things were not looking good for the Olympians until Hermes and Pan managed to find and return Zeus's tendons to him. Zeus then chased Typhon, continuously bombarding it with thunderbolts, eventually killing it. Its corpse lies beneath Mount Etna in Sicily.

Hera, his sister, became the wife of Zeus. Hera proved extremely jealous of Zeus's many lovers. Zeus's seduction techniques were frequently deceitful. He apparently 'persuaded' Hera into becoming his wife by disguising himself as a fluffed-up, feeble-looking cuckoo. Hera pitied the frail-looking bird and put it against her breast. At that moment Zeus transformed himself back and had sex with Hera. She became

Fig. 9.2 Oak trees, being associated with Zeus, are also linked to Jupiter. (Illustration by the author)

his wife, and their honeymoon lasted for 300 years! Despite three centuries of passion, Zeus became an adulterer and rapist, as the mythology concerning the planet Jupiter's moons reveals.

Zeus, though he was the supreme god and the most powerful, still angered quickly. Hera, when she became jealous, had to be careful about how she disposed of his lovers. She even had to resort to occasionally borrowing Aphrodite's enchanted belt (which made everyone fall in love with its wearer) to calm him and reignite his passion. She and the other gods got really fed up with his temper, so Hera decided to try and overthrow him. The gods got together and bound the powerful Zeus, making sure that his potentially fatal thunderbolts were where he could not reach them. With help Zeus eventually escaped the bonds and decided to make an example of his treacherous wife, so he hung her from his sky with weights on her wrists and ankles—very undignified!

When Lycaon's sons served him soup containing the entrails of one of their own brothers, Zeus became so angry that he wanted to annihilate the entire human race. He turned the sons of Lycaon into wolves, perhaps the forerunners of the werewolf. Then he caused an enormous, Earth-consuming flood, during which most of the human race were drowned.

To the ancient Greeks, Zeus was so important that in 438 B.C. the sculptor Pheidas carved an enormous 60-foot statue of the supreme god. This was regarded as one of the Seven Wonders of the World. The eagle is an attribute of the god, as he sent two eagles from opposite ends of Earth. They met at Delphi in Greece. This became the home of the famous oracle of Apollo. Purple is the color of Jupiter, as king of the gods, a color that today is still associated with royalty. The red eye or storm on the planet Jupiter is associated with the blood of Christ. Zeus can also be represented by oak trees and more obviously by bolts of lightning (Fig. 9.2).

Even the expression "By Jove!" is about Jupiter, used today without people even thinking of its origin. When the weather changes for the worse on Mount Olympus, Greeks even today say that Zeus is in a bad mood. They still retain a degree of respect for the ancient deity.

In some stories, Alexander the Great (356–323 B.C.) is said to have been one of Zeus's many offspring. Alexander himself is believed to have visited the oracle at Siwa in the middle of the desert to ask if he was the son of Zeus. The oracle told him that he was. This answer undoubtedly influenced his behavior and actions, as he would have believed that he was semi-divine and therefore virtually invincible. After his death, a coin was produced featuring Alexander wearing the ram's horns of Zeus, linking him to the supreme god.

To the ancient Greeks, the Titans were symbolic of chaos and disorder in contrast to Zeus, the personification of the birth of order. Zeus sired many offspring, including the heroes Aeneas and Hector, who involved him in the Trojan War. The Roman god Jupiter was also associated with Baal, a god mentioned in the Bible. Jupiter/Zeus was linked to gods from other cultures, including the Armenian chief god Aramazd, the Norse god of thunder; Thor, the Germanic lord of thunder and lightning; Thunor or Donar, the ancestor of Thor; the Germanic sky father god Tiwaz; and Tyr, the Scandinavian equivalent. Even the Celtic god Taranis, meaning "thunderer," was sometimes identified with Jupiter.

The Moons of Jupiter

The planet Jupiter has at least 60 moons, most of which have been designated with a name, usually that of a lover of the supreme deity.

Io

Io is the Galilean moon nearest to Jupiter (Fig. 9.3). It is the most volcanic body in our Solar System and resembles a spherical pizza. Its pocked surface displays a mixture of autumnal hues of yellow, brown and orange, due to the element sulfur.

Io is too small to be able to support internal heat of its own, so it must be the phenomenal power of Jupiter's gravitational pull that powers the volcanoes on this moon. This vigorous energy from Jupiter is constantly expanding and contracting Io in a gravitational tug of war. Io's first recorded plume was given the name Pele, after the tempestuous Hawaiian fire goddess.

Io was one of Zeus's many lovers. She was the daughter of the river god Inachus. She was also a virgin priestess at Hera's temple at Argos. Zeus's attraction to her was bound to seriously annoy Zeus's wife Hera. Zeus was not in the least bit discouraged by this. After all, he was the supreme leader of the gods and could do exactly what he wanted. He got Io expelled from Argos to make her more accessible

Fig. 9.3 A view of Galilean satellites. (Image courtesy of NASA)

Fig. 9.4 Zeus, or Jupiter, seduced Io as a cow. (Illustration by the author)

to him. In one version, Zeus turned Io into a beautiful, young, snow-white cow (Fig. 9.4). He would have had sex with Io at once, but Hera, ever-suspicious, guessed what Zeus was up to and sent a fly to torture poor Io, to prevent her from standing still. Zeus, not one to give up easily, eventually had sex with Io on a cloud over Egypt, where he transformed her back to her original, human form. Amazingly, Io was forgiven by the usually schadenfreuden Hera.

In another version of this myth, Hera made Zeus give Io in cow form to her, and set Argos, Hera's 100-eyed watchman, to guard the beautiful cow. Poor Io tried to call her father and sisters to help her but could only make a mooing sound. She eventually wrote her name in the earth with her hoof. They guessed that the cow

Fig 9.5 After Hermes killed Hera's servant Argos, myth claims that Hera immortalized his many eyes in the design of a peacock's tail. (Illustration by the author)

was Io from this. Zeus learned about this and enlisted the help of Hermes. Hermes visited Argos and told him long, boring stories to try and make all 100 eyes close in sleep. The stories only partially worked, so Hermes shook a bunch of poppies over his head, sending him to sleep with all of his eyes finally tightly shut. Then, rather unnecessarily, Hermes killed poor sleeping Argos and led Io back to Zeus, where he changed her back into her original form. Hera was upset at finding her faithful servant dead and placed his 100 eyes in the peacock's tail, where they remain for us all to see (Fig. 9.5).

Europa

This is the smallest of the Galilean moons, with an icy crust encasing its volcanic surface. Surface temperatures are −160 °C. It is believed that the surface is extremely dense ice floating on the top of an expanse of water rather like that on Earth in the Arctic and Antarctic regions. Europa's surface bears hardly any evidence of meteor impacts, so it must be able to replenish itself somehow. Europa is a celestial body that could possibly contain some kind of life. Scientists are studying microbial life in a Yorkshire mine, where conditions are dark, hot, dusty and salty to learn about potential life on Europa.

In mythology, Europa was the daughter of Telephassa and King Agenor of Tyre in Phoenicia. Zeus, this time disguised himself as a beautiful white bull, swam to the island of Crete with King Agenor's daughter, Europa, riding on his back. They had sex and together produced three sons, Minos, Rhadamanthys and Sarpedon, before she was married off to a local ruler, Asterius. Asterius was compensated by Zeus for his having taken Europa's virginity. He gave Asterius a giant man made out of bronze metal called Talos for the purpose of defending his territory. In a variation of this myth, Europa was a beautiful young woman picking flowers when Zeus, transformed into a bull, enticed her with a mouth full of pretty crocuses. Then he kidnapped Europa and took her to Crete.

Europa appears in the Periodic Table as Europium, with the atomic number of 63 and the symbol 'EU.'

Ganymede

Ganymede is the biggest moon in our entire Solar System, even larger than the planet Mercury and the dwarf planet Pluto. It is the only moon that hints at a possible iron core. It has a blotchy surface in various grayish hues, textured and cratered from meteorite impacts. Its orbit is circular. It is also conjectured that, as its surface contains both older and younger rocky and icy parts, the resurfacing could indicate the presence of tectonic plates, like those that exist on Earth.

In mythology, Ganymede was the son of Tros, king of Phrygia. He was such an incredibly beautiful young man that Zeus, disguised as an eagle, stole him, taking him to his home on Mount Olympus to become his cupbearer. In one legend, Ganymede drank a potion made from the flower Tansy so that he could live with the gods forever. In a variation of the same legend, Ganymede became one of the Zeus's many lovers and gained immortality as the constellation Aquarius, the water-carrier.

Callisto

The second largest and furthest away from the parent planet of the four Galilean moons, this moon is only fractionally smaller than Mercury, which part of its surface resembles, suggesting that they are of a similar age. As the planet Jupiter attracts comets, often fragmenting them, Callisto and other moons are right in their way, hence its dimpled and battered cratered surface. It does not generate its own heat, as it is too small. However, it does have a magnetic field, and the most likely

Fig. 9.6 Zeus transformed the beautiful Callisto into a bear to protect her from Hera's wrath; Callisto and her son were later made constellations. (Illustration by the author)

explanation that scientists have come up with for this is that it is created as Callisto travels through Jupiter's amazingly strong magnetic field, but this is still just a theory.

In mythology Callisto, meaning "the most beautiful," was either an Arcadian princess or a wood nymph. Zeus noticed her and had sex with her. To protect her from Hera's anger and jealousy, he turned her into a bear (Fig. 9.6). They produced a human son, Arcas, who was afraid of his bear of a mother. To protect the frightened Arcas, Zeus transported them both up to the heavens. Arcas is in the constellation Boötes, with Callisto part of a group of circumpolar stars. In a variation of this myth, Zeus metamorphosed into Artemis, virgin goddess of the hunt, women and the wild, in order to sleep with the gorgeous Callisto.

From Arcas the Bear comes the name Arctos for the Arctic, which the Greeks knew about.

ASGARD. This is an impact basin on Callisto. In Norse mythology, Asgard is the realm of the Aesir, who are the younger, stronger branch of the Norse pantheon. Asgard is supposed to be in the center of the universe and is only accessible by the rainbow bridge called Bifrost. Asgard contains regions and mansions that include Valhalla.

VALHALLA. This is an impact basin on Callisto. The Valhalla impact basin is 600 km (375 miles) across, framed by concentric ridges extending almost a third of the way around Callisto. It is very likely to have been the result of a collision with a comet.

Valhalla is a hall within Asgard, the home of the gods. It is the hall of Odin/ Woden, the supreme god of the Aesir dynasty and the god of war. His battle maidens, the Valkyries, selected the bravest and boldest of those slain in battle, whose souls were taken to Valhalla to eat, drink and fight every day until the final battle of Ragnarok, when their gods and frost giants will all die and the world will end.

Amalthea

The moon Amelthea is less than 250,000 km (155,000 miles) from Jupiter.

In mythology, Amalthea was *not* a lover of Zeus. She was a faithful goat-nymph who brought up Zeus on the island of Crete. She died when Zeus was almost an adult. Zeus had her skin made into a magically strong shield.

Himalia

This moon was named for a nymph with whom Zeus sired three sons.

Elara

Mother by Zeus of the giant Tityus. Tityus tried to rape Leto, the mother of the twins Artemis and Apollo. The twins killed him for this.

Pasiphae

This moon is for the wife of King Minos. Poseidon made her fall in love with a bull. She then became the mother of the infamous Minotaur (meaning "Mino's bull"). The Minotaur was a monster with the body of a man and the head of a bull. According to Ovid, Pasiphae climbed into a wooden model of a cow made by Daedalus in order to have sex with the bull.

Sinope

Named for the daughter of the river god Asopus. Zeus wanted her so much that he promised to give her any gift that she desired. She cleverly chose virginity as her gift, and deeply disappointed the lustful Zeus.

Lysithea

Lover of Zeus and daughter of Oceanus.

Carme

Carme was a nymph who helped with the grain harvest. She was also an attendant of Artemis. Zeus sired a daughter, Britomartis, by her.

Ananke

Ananke and Zeus produced a daughter, Adrastea, who later became an allocator of rewards and punishments.

Leda

This moon is named for Leda, the daughter of King Thestius of Aetolia. She was married to King Tyndareos of Sparta. Zeus wouldn't let a mere trifle like marriage get in his way, so he disguised himself as a swan and had sex with Leda (Fig. 9.7). Leda produced two eggs a while later. From one egg hatched the twins Caster and Polydeuces, from the other hatched the infamous Helen and the formidable Clytemnestra, who murdered her adulterous husband.

Fig. 9.7 In yet another escapade, Zeus raped Leda in the form of a swan. (Illustration by the author)

Metis

This moon is less than 250,000 km (155,000 miles) from Jupiter. In mythology, Metis was an intelligent sea nymph. She transformed herself into several different forms in order to avoid Zeus's amorous advances. However, Zeus was not one to take no for an answer and had sex with her. She became pregnant and Gaia, the Earth goddess, predicted that she would have two children—a girl and then a boy who would become ruler of heaven. Anxious to avoid being overthrown by his own son, Zeus swallowed Metis and her unborn female child, gaining Metis's wisdom in the process. The child was eventually born from Zeus's head and became the virgin goddess of wisdom, Athena.

Adrastea/Adrestis/Adrastus

Mentioned in Homer's *Iliad* as a city northeast of Troy. Adrastea is likely to have been a local nymph of the city.

Thebe

Less than 250,000 km (155,000 miles) from Jupiter, this moon is named after the city of Thebe, mentioned in Homer's *Iliad* and destroyed by Achilles. Thebe and her twin sister, Aegina, were daughters of the river god Asopus. Thebe married Zethus.

Themisto

A tiny little moon just 8 km (5 miles) across, Themisto was discovered in 1975 but then went missing and was not seen again until the year 2000.

In mythology is a tragic tale of a jealous second wife, Athama, who was originally married to Ino, by whom they had twin sons called Learchus and Melicertes. One day, Ino went out hunting. She failed to return, and her clothing was found spattered with her blood. Ino was assumed to be dead. After a time of mourning passed, Athama married Themisto, who also bore him twin sons. Unfortunately, it was at this point that Athama discovered that Ino was still alive. He brought her into his house as a nursemaid. Themisto then worked out Ino's true identity and was jealous and angry. She got Ino to prepare white clothing for her twin boys and black for Ino's. Then, Themisto instructed her guards to kills the black-clothed twins and to save the others. Ino was suspicious when told not to dress the sets of twins and swapped their clothing around. This resulted in the murder of Themisto's twins. Ino managed to escape, but Athama went mad when he learned of this.

Carpo/Carpho

One of the Hours or Horae, personifications of the seasons. Carpo represented vegetation in one myth; in another Carpo represented winter.

Europie

Europie was another of the Hours or Horae. She was Zeus's daughter by Themis. It is possible that Europie could be another name of Auro, meaning "growth."

Thelxinoe

Supposedly this was one of the original three Muses, figures that represented inspiration in different areas.

Euanthe/Euanthes

This is the son of Dionysus and Ariadne. Rhadamanthys, the son of Zeus and Europa, gave some land in Maroneia to Euanthes.

Helike/Helice

Mentioned in Homer's *Iliad* in Agamemnon's kingdom, this city was sacred to Poseidon. The four classes of farmer, craftsman, soldier and priest in Athens were allocated the names of the sons of Helice and Ion.

Orthosie

This moon was named after a minor goddess of prosperity.

Iocaste/Epicaste

The name means "shining moon." Iocaste was the mother of Oedipus. She and her husband abandoned their son as a baby, as it had been prophesied that he would sleep with his mother and would murder his father. Oedipus was found and brought up by others. He did kill his father and had sex with Iocaste, his mother, although they were unaware of their true relationship at the time. Iocaste hung herself when she discovered the truth.

Praxidike

This was the Greek goddess of carrying out justice.

Harpalyke/Harpalyce

Clymenus had sex with his daughter, Harpalyce. He then married her off to Alastor but decided to take her back after she was married. Harpalyce was angry and wanted revenge, so she murdered their baby son, cooked its corpse and laid it out for her father to see. Clymenus, naturally, was distraught and hanged himself. Harpalyce was transformed into a bird of prey.

Mneme

This was one of the three original Muses, daughters of Zeus and Mnemosyne.

Hermippe

This was the mother of Orchomenus by Zeus.

Thyone

Dionysus descended from heaven, via Lerna in Argos, to Tartarus. He bribed Persephone with myrtle to release his dead mother, Semele. Semele ascended with him, but Dionysus changed her name to Thyone so that the other ghosts would not be upset and envious. In a variation of this story, Semele was transported up into the sky with the stars as Thyone.

Herse

Cecrops, son of Gaia, married Agraulos. Together they had three daughters: Aglauros, Herse and Pandrosos. Hermes/Mercury had fallen in love with and lusted after Herse, who was the youngest and most beautiful of the sisters. He bribed Aglauros with gold to arrange a meeting with her. Aglauros decided to take the gold but refused to help. She should have known better than to try to deceive a god.

Hermes was absolutely furious and turned the greedy Aglauros to stone and had sex with Herse, who gave him two sons.

Aitne

This moon is named for one of the many nymphs that Zeus slept with.

Kale/Cale

Tiresias had the unfortunate and unenviable task of deciding who was the most beautiful out of the goddess of love, Aphrodite, and the three Charities: Pasithea, Cale and Europhasyne. Cale was awarded the prize and Aphrodite, angry and jealous, turned Tiresias into an old woman. Cale took him to Crete with her and rewarded him wonderful hair.

Taygete

One of the seven Pleiades. The goddess Artemis had temporarily transformed Taygete into a female deer to elude the lustful Zeus (Fig. 9.8). Zeus, ever persistent, raped her. She gave birth to Lacedaemon, the founder of Sparta. Taygete was utterly distraught and hanged herself on the summit of Mount Amyclaeus, renamed Mount Taygetus in her memory.

Chaldene

This moon is named for another of Zeus's lovers and mother of Solymus.

Fig. 9.8 Zeus struck yet again in the rape of Taygete, transformed by Artemis into a deer. (Illustration by the author)

Erinome

This was a daughter of Poseidon and mother by Zeus to the three Graces.

Aoede

This was a daughter of Zeus and the Muse of Song. According to the Greek people, the Muses evolved into fairies, which some people still believe to exist.

Kallichore

This is another Muse and daughter of Zeus.

Kalyke/Calyce

A nymph. Calyce and Poseidon had a son, Cycnus. Zeus slept with Calyce, too, and Endymion, who became Selene, the Moon's lover.

Kallirrhoe/Callirrhoe/Callirhoe

This moon was named for the daughter of the river god Achelous. Callirrhoe married Alcmaen. They had two sons. She was also the lover of Chrysaor, who sprang from the severed head of the Gorgon named Medusa. Together they produced the three-headed Geryon. In Ovid's *Metamorphoses*, Callirrhoe asked Jupiter to make her children into adults in order to avenge the death of her husband.

Eurydome/Eurynome

She was the goddess of all things. Along with Thetis, she rescued Hephaestus, the smith-god whose mother, Hera, threw him into the sea from the top of Mount Olympus. They kept him in an underwater home, where he learned to become a metal smith. Eurynome was the mother of the Graces by Zeus. She was also Leucothoe's mother. Leucothoe was so beautiful the Sun fell in love with her.

Kore/Core/Semele/Persephone

This moon was named for the daughter of Zeus and Demeter. Pluto fell in love with Persephone. He asked the permission of Zeus to marry her. Zeus refused to give him an answer, so Pluto stole Persephone and took her to the Underworld. Core, Persephone and Hecate were a goddess in triad. Core symbolized the new, green corn, Persephone the ripe corn and Hecate the harvested corn.

Cyllene

This is a nymph who acted as nurse to Hermes/Mercury.

Eukelade

This daughter of Zeus was supposed to be one of the Muses.

Hegemone

This is one of the two Charities of Athens. Hegemone can be identified with winter.

Arche

This is one of the original Muses—daughter of Zeus and Mnemosyne.

Isonoe

This lover of Zeus was one of the fifty daughters of King Danos of Argos; the Danaides all murdered their husbands on their wedding night.

Sponde

This was one of the Horae or Hours, daughters of Zeus and Themis. The Hours were goddesses of the seasons before being associated with divisions of time.

Autonoe

Daughter of Cadmus and Harmonia, this was the wife of Aristaeus. They had two children, Macris and Actaeon. Actaeon saw the virgin goddess Artemis/Diana bathing naked. She was furious and in retaliation turned him into a stag. He was hunted and torn to pieces by his own dogs.

Megaclite

This was yet another lover of Zeus who later became queen of Crete.

Pasithee/Pasithea

The god of sleep wanted to marry Pasithee, out of all of the Graces. Hera bribed the god by promising him Pasithee so that he would cause Zeus to fall asleep.

Jupiter/Zeus sired many more offspring. The most important include the Horae or Hours and the Fates or Moirae by Themis, a Titaness whose name means righteousness; the Muses by Mnemosyne, another Titaness; Argos, who founded the city that bore his name by Niobe, a mortal woman; and Herakles/Hercules by Alkmene, a princess of Argos. Zeus did not play fair with Alkmene and appeared in the form of her husband, Amphitryon. The supreme deity also appeared as a shower of gold to Danae, another princess of Argos by whom he begot Perseus. In some versions of mythology, Herakles is supposed to be descended from Perseus, which would still have made him a relative of Zeus.

The Ice Giant Planets Beyond Jupiter

Astronomers continue to discover new information about the Solar System. However, they have difficulty in explaining the large size of the outer, ice-giant planets. The 'jumping Jupiter' theory has been around for a while, namely that Jupiter must have originally had a position in the Solar System closer to the Sun.

Astronomers continue to try to solve cosmological conundrums. Most recent scientific conjecture suggests that our Solar System once included an extra planet in an attempt to explain the large size of the two outer ice giant planets.

It is theorized that when our Solar System was formed, the planets were initially much closer together but then ventured further afield at a later time, with the planets with the strongest gravitational pull fighting for their places. This would probably have resulted in one of the planets being expelled from our Solar System. However, at the moment, without more concrete proof, this remains only a theory.

See Chap. 11 for more on the outer ice giants.

Chapter 10

Saturn, a Gaseous Beauty with Many Moons

Saturn is an incredibly beautiful planet with a somewhat ethereal, incorporeal appearance (Fig. 10.1). Some merely regard it is a lesser Jupiter, but Saturn is of huge scientific interest in its own right.

The Ringed Planet Superstar

Although all the gas giants are encircled by rings, Saturn's are the most curious and the brightest in our Solar System. They are the most visible, too. Some at first thought that Saturn had handles, and when the Italian astronomer Galileo first saw Saturn with its rings, he thought that it had ears! It took until 1665 for the Dutch astronomer Christiaan Huygens to work out that Saturn was encircled by a ring, and until the Victorian era before it was revealed by the Scottish scientist James Clerk Maxwell that the rings were actually composed of millions and millions of minute individual particles. It is believed that Saturn's rings are actually remnants of either a tiny moon or a comet that has a circular orbit.

Solid rings could not have formed, as they would have been destroyed by the planet's gravitational pull. Saturn's rings are made up of hundreds of individual ringlets, some only a few meters in thickness. These are ice and dust particles.

The planet is extremely light, with a lower density than any other planet in our Solar System and a less powerful gravitational pull than that of Jupiter. Saturn itself is a ball of gas, rather like Jupiter composed predominately of hydrogen and helium. Like Jupiter, Saturn has violent storms, some the size of Earth. Saturn's own internal energy is the power behind these winds.

© Springer Science+Business Media New York 2015
R. Alexander, *Myths, Symbols and Legends of Solar System Bodies*, The Patrick Moore Practical Astronomy Series, DOI 10.1007/978-1-4614-7067-0_10

Fig. 10.1 The majestic rings of Saturn set it apart. (Image courtesy of NASA)

Saturn/Cronus in Mythology

Saturn is the second largest planet in our Solar System; this is perhaps the reason why it is named after Jupiter's father in mythology, as he was also a leader of a dynasty of gods—the Titans. Saturn was originally named after an ancient Italian god of corn who gradually became identified with the Greek god Cronus.

Cronus/Saturn is perhaps best known for the castration of his father, Uranus. Mother Earth, or Gaia, wanted retaliation for Uranus exiling the Cyclopes to Tartarus, part of the Underworld, so she conspired with the Titan gods led by Cronus/Saturn to overthrow Uranus. She provided that famous sickle, which in some versions of the myth is made of serrated stone, rather like a saw.

Apparently, Cronus/Saturn held his father's reproductive organs with his left hand while doing the cutting or sawing, depending on how the weapon is imagined, eventually removing them with his dominant, right hand—a gruesome image. He flung the bloody, severed testicles and penis along with the cruel weapon into the sea.

Cronus/Saturn retained his leadership of the Titan gods until he was overthrown by his own son, Zeus/Jupiter. As told earlier, while bleeding to death, Uranus predicted that his son would, in turn, be overthrown by one of his sons. To prevent this from happening, Cronus swallowed all of his children, even the females, just to make sure that the prediction had no chance of ever coming true. Rhea, Cronus's wife, was not exactly happy at seeing her children being swallowed. She gave birth to the youngest child, Zeus, in darkness in an obscure place. The baby was brought up by the nymphs Adrastea and Amalthea. Rhea fooled her husband by substituting a stone wrapped in swaddling for the baby Zeus. Cronus/Saturn swallowed the stone.

When he became an adult, Zeus was able to smuggle a magic potion given by his mother, Rhea, into Cronus's drink. This forced him to throw up the stone, and

Fig. 10.2 Ravens are associated with Saturn. (Illustration by the author)

then his brothers and sisters, collectively known as the Olympian gods. Then, a 10-year war began, the Olympians led by Zeus and the Titans, not led by Cronus, who was considered too old, but by the enormous Titan called Atlas.

Cronus/Saturn was eventually overthrown by underhanded means. The three most dominant Olympians worked as a team to overthrow their father with the help of the presents given to them by the Cyclopses, the one-eyed giants. Hades/Pluto wore his helmet of invisibility and stole his father's weapons, leaving him defenseless and vulnerable, while Poseidon/Neptune took away Cronus's/Saturn's attention from Zeus with her trident so that Zeus could hurl a thunderbolt at him. Cronus then was forgotten. In another version, Zeus gave his parents the Isles of the Blessed to reside over, a more peaceful ending. In Orphic creation mythology, Cronus was the offspring of Nyx (Night) and Phanes, the androgynous ancestral god born from an egg that Chronus created.

Saturn is generally depicted as an old man with a gray beard and a scythe, rather like the Grim Reaper. Like the Grim Reaper, who collects people when their time on Earth is up, Saturn is associated with time and Old Father Time. The color black or blue-black is the color of chaos and of Saturn. The metal lead is associated with Saturn, too. Ravens are supposed to have some similar characteristics to Saturn, including as creatures of superstition (Fig. 10.2). The sight of a raven can reveal the coming bad weather, and it is believed that when ravens leave their normal nesting place, this foretells of famine, death and calamity. Perhaps this is why the ravens at the Tower of London all have their wings clipped (Fig. 10.3).

The planet Saturn is best known, perhaps, by the ancient Roman mid-winter festival that commemorated it, called Saturnalia. This occurred around December 25, when the Romans used to give presents and celebrate the victory of light over darkness. This festival was adopted by the Christians, and it became Christmas. The dead return during the 12 nights of the Saturnalia period. The Christians turned this into the 12 days of Christmas.

Fig. 10.3 A raven at the Tower of London. (Illustration by the author)

Saturn's Moons

Saturn is an exciting planet with over 60 known moons, ten of which are, at the time of writing, unnamed. Some of Saturn's moons deviate from the traditions of being named after characters drawn from Greco-Roman myth, instead using names of characters from legends created by other peoples, including Norse and Inuit.

Titan

This is a giant moon, Saturn's largest, hence the name (Fig. 10.4). This moon is, in fact, larger than the planet Mercury. It is also featured as the metal titanium in the Periodic Table, with the atomic number 22 and the symbol 'Ti.' Titan measures 5,100 km (3,200 miles) in diameter and is the second largest moon in our Solar System, the largest being Jupiter's Galilean moon, Ganymede.

Titan is a curious moon, the only satellite with a considerable atmosphere, which is a strange orangish-yellow color. This atmosphere consists mainly of nitrogen plus a very small amount of methane. As Earth has a hydrological cycle, Titan has a methane cycle, with rain and lakes consisting of methane, and methane from

Fig. 10.4 Saturn's methane moon Titan. (Image courtesy of NASA)

volcanoes has been observed. Its surface is hidden by a veil of citrus-colored atmosphere, but some relatively new impact craters have been noted. There is no liquid water on Titan, so it is unlikely to harbor life. However, it is scientifically interesting, as it is covered in liquid organics.

The Titans were a dynasty of deities, the offspring of Gaia and Uranus, led at first by Cronus, then by the mighty Atlas. They represented primordial chaos or disarray, in contrast to Zeus and the Olympian gods, who were symbols of order. Zeus imprisoned most of them in Tartarus. The Titans can be seen as power-hungry. The 10-year battle between the Olympians and the Titan gods impacted the whole world.

The word Titan itself has become somewhat infamous, due to the tragic sinking of the enormous 45,000 tonne ocean liner, the RMS *Titanic*. Three different companies—Cunard, the Hamburg-Amenka and the White Star—competed to rule the seas by building bigger and bigger vessels. The *Titanic* was the culmination of this, the largest steamer in the world, supposedly unsinkable. The *Titanic* collided with an iceberg, though, resulting in the deaths of over 1,500 people, a tragedy that shook the whole world. Most of the dead froze to death rather than drowning, and there were not enough lifeboats for everyone, despite there being more than required by the regulations of the day.

There is much superstition surrounding this terrible event, including a curse from an Egyptian mummy, a gypsy's prophesy and premonitions from individual passengers. Under the pressure of the enormous volume of below-freezing water, the ship tore herself apart, metal shrieking, rivets shattering. The ship, due to its size, was comparable to a monster, but the North Atlantic Ocean was far more powerful and vanquished her. The once-great luxurious liner was reduced to a mere wreck on the seabed, so far down that it took over 80 years for the technology and resources to become available to enable the wreck to be revisited. Many regarded it as somewhat hubristic when the White Star line boasted that their ship was 'unsinkable.' The *Titanic* has become a powerful symbol of doom, death and destruction, even of hubris and nemesis.

Iapetus

This is a very interesting, spherical moon with a raised ring encircling its equator. It is gravitationally locked, with one of its faces permanently watching its parent planet. Half of Iapetus's surface is reflective and icy, the other half dark and light-absorbing. Its surface is also covered in craters. The moon has a mountain range on its equator.

In mythology, Iapetus was one of the six male Titans, the father of Prometheus by the nymph Clymene, among others.

Rhea

A spherical, heavily battered and bruised ball of ice, slightly larger than Iapetus is Rhea. Rhea was one of the Titan goddesses and was both the wife and mother of Cronus/Saturn. Her parents were Uranus and Gaia. She was the mother of Hestia, Demeter, Hera, Hades, Poseidon and Zeus. She saved her youngest son, Zeus, from being swallowed by her husband by substituting for the baby a stone covered in swaddling.

Dione

Another heavily cratered, battered and bruised spherical moon. Dione shares her orbit with two smaller moons, the Trojan moons Helene and Polydeuces. Dione is similar in size to Tethys (see below) and is Saturn's fourth largest moon. Its darker hemisphere has enormous cliffs of ice on it.

In some mythologies, Dione was the mother of Aphrodite, while in others she is a goddess of the oak tree. She is quite possibly the mother of Dionysus, with Zeus being the father.

Tethys

Similar in size to Dione but slightly smaller in diameter, this is another very heavily cratered and battered world of ice that is dominated by its largest crater, named Odyssey.

Today, the word odyssey merely means some kind of journey, after the epic journey of Homer's hero Odysseus in his poem *The Odyssey*. Odysseus fought in the Trojan War but incurred the wrath of Poseidon, who tried to destroy him and caused him to feel great despair. A usually quick journey lasted over 10 years due to Odysseus's many adventures along the way. Eventually he did find his way home to his loyal and patient wife, Penelope, and his son, Telemachus.

Tethys' parents were Gaia, the Earth mother, and Uranus. She became the wife of Oceanus, who ruled over the sea. They became parents of 3,000 rivers and the same number of sea nymphs, called the Oceanides. Tethys was a moon goddess as well in ancient mythology. She was also the grandmother of Phaethon, who fool-ishly drove the Sun-god Apollo's chariot at a ridiculously fast and irresponsible speed and ended up dying. Finally, Tethys was the name of a vast ocean millions of years ago in Earth's history. It no longer exists, as the continent of Eurasia formed from the demise of the mighty ocean.

Mimas

This innermost spherically shaped moon of Saturn has more than a third of its sur-face occupied by the enormous Herschel crater, the remaining evidence of a par-ticularly colossal collision that almost completely shattered the moon. The Herschel crater measures 140 km (80 miles) across. Mimas looks like the Death Star in the *Star Wars* films, with its dominant crater, and has a mountain range on its equator.

In Greek mythology, Mimas was a giant. Hephaetus, the smith-god, flung a ladle of scalding-hot metal at him during the battle between the Olympian gods, led by Zeus and the giants in retaliation for Zeus imprisoning the Titan gods in Tartarus. (Figure 10.5)

Enceladus

This is the brightest and most reflective moon yet known in our Solar System (Fig. 10.6). Its surface is a dazzling shade of white and is icy. It reflects back almost all of the sunlight that reaches it. It also has a flimsy atmosphere surrounding it.

Enceladus, despite being a tiny moon, may actually be able to harbor life. There is warmth at the 'tiger stripes' from many tall geysers that spit out water vapor containing organic material

Fig. 10.5 The moon Mimas is notable for its large Hershel crater. (Image courtesy of NASA)

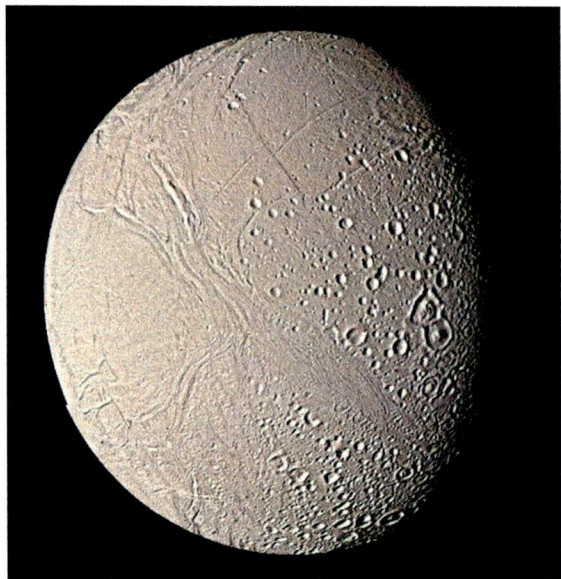

Fig. 10.6 The icy surface of Enceladus. (Image courtesy of NASA)

In Greek mythology, Enceladus was the leader and therefore the largest of the 100-armed giants. Children of Uranus and Gaia, they attacked the Olympian gods in what was known as the Giants' Revolt. Zeus defeated him in battle and flung him down to Earth at Macedonia and was supposed to have thrown Mount Etna on top of him. The vanquished gargantuan monster, Typhon, is supposedly buried beneath Mount Etna in Sicily. Enceladus, one of Typhon's brothers, is said to produce the flames of the volcano with his angry breath.

Hyperion

Hyperion looks like a natural sponge! It is conjectured that this sponge-resembling object could easily be a piece of a moon that was once much bigger but was obliterated by a collision. Hyperion was discovered in 1848.

In Greek mythology, Hyperion was a Titan god, the father of Eos, goddess of the dawn, Helios, a Sun god, and Selene, the goddess of the Moon, by Theia. Hyperion had seven herds of cows and many sheep grazing on the island of Sicily. When Odysseus landed at Sicily he forced his men to swear an oath not to steal any of Hyperion's cows. The men were stranded on Sicily and were therefore hungry, so they were unable to resist killing a few cows. This act infuriated Zeus and Poseidon, the latter of which conjured up a violent storm, leaving all of Odysseus's men dead.

Phoebe

A small moon at only 214 km (133 miles) in diameter, Phoebe is too tiny for gravity to transform it to a perfect sphere. It is generally thought to be an object attracted by Saturn's gravitational field originating from the Kuiper Belt. Phoebe is the nearest of the irregular satellites to the parent planet. It travels in the wrong direction around Saturn and appears to be coated with a red, possibly dusty substance. Phoebe's surface, like that of Earth's own Moon, has been formed by many collisions.

Phoebe was one of six female Titans. The name means "bright moon." She was the mother of Leto and grandmother of the twins Apollo and Artemis.

Epimetheus

Another small moon at only 122 km (75 miles) in diameter, this resembles a huge, crudely cut jewel with weird facets. Epimetheus shares its orbit with the moon Janus, suggesting that both moons are two parts of a once much larger one that broke up.

Epimetheus was a Titan god, a son of Gaia, the Earth mother, and Uranus, the sky father. His brothers included Hyperion, Atlas and Prometheus. His name means "hindsight." Zeus sent him Pandora, a beautiful woman made by Hephaestus, the smith-god. Despite being warned not to accept this gift, he eventually did and married Pandora.

Janus

A slightly larger irregular moon than Epimetheus, with whom it co-orbits, Janus was an ancient Italian god symbolizing origins and sources. He was a two-faced god who watched over life's dualities. The month of January is associated with Janus, which is a time when people look to the past and to the future. Janus, with his additional two eyes literally at the back of his head, was able to catch the nymph Carna, who usually enjoyed teasing potential lovers before running off. Janus had a temple dedicated to him in Rome with two gates, always open during war but closed when at peace. To call someone a Janus-face is to accuse them of duplicity and hypocrisy.

Atlas

This non-spherical moon was discovered in 1980.

Atlas was a Titan god, selected to lead this dynasty of deities against the Olympian gods led by Zeus. He was chosen to lead for two reasons. The first was that Cronus/Saturn was believed to be too old, and the second was his mighty strength and stature. Although Zeus banished the other male Titans to Tartarus, Atlas was made an example of. His eternal punishment was to hold up the heavens, or in some versions, Earth itself. His parents were Iapetus and the nymph or Oceanid called Clymene. When Phaethon drove Apollo's chariot so recklessly that he ended up dead, Atlas's shoulders were straining under his heavy load.

In another myth the Greek hero Perseus pretended to give Atlas a present and produced something out of his bag. The present was none other than the head of the Gorgon Medusa. Consequently, Atlas was turned into a huge mountain, growing in stature until the very stars appeared to rest on its peaks. Atlas was temporarily relieved from his sky-holding duties by the hero Heracles, who offered to take over temporarily while Atlas fetched him the apples of the Hesperides. Atlas even offered to deliver these apples for him. Heracles pretended to agree on the condition that Atlas took the world for a moment. Atlas took back the world and then realized, too late, that he had been tricked by the cunning Heracles. In another myth, Atlas's massively broad shoulders supported an enormous column that was used to separate Earth and the sky.

The phrase 'the weight of the world on your shoulders' comes from the story of Atlas.

Prometheus

Prometheus, as well as Pandora, are regarded as shepherd moons. Prometheus is irregular in shape.

Prometheus is a son of Iapetus and brother to Epimetheus. He is the protagonist in Aeschylus's play *Prometheus Bound* and appears as a champion of the human race. This is perhaps unsurprising for the god that actually fashioned humankind out of clay. Zeus had intended to kill off all humans and to replace them with another race. Prometheus gave humankind gifts that would enable them to survive, blatantly challenging the supreme authority of Zeus, hence his terrible punishment. He stole fire from the Olympian gods and gave it to humankind plus the knowledge of how to use it. He also gave them hope, carpentry, skills and science, taught them how to write down words, how to heal with herbs, how to make craft to transport them over water, how to interpret prophecies, how to use animals to carry heavy loads and how to ride them. He also gave humans precious metals hidden deep within Earth, including iron, silver and gold. Finally, he provided humans with the gifts of reason and how to apply experience and knowledge, how to build houses and how to read the stars.

Prometheus' punishment was truly terrible. He was an example to other gods who felt they could challenge the authority of the almighty Zeus. Hephaestus, the smith-god, reluctantly made bronze and steel bands to secure Prometheus with. Hephaestus did not enjoy inflicting a terrible punishment upon a god of his own race, but was too afraid of Zeus to refuse. Prometheus was also secured with a diamond chain to the peak of a mountain, where he would be subjected to pain every hour of the day. Every night, he was tortured further by having his liver devoured by an eagle, a symbol of Zeus's power. The liver grew back each day and was eaten again each night. This supposedly eternal punishment was stopped when Prometheus was rescued by the courageous Heracles, who killed the powerful eagle and unchained him.

According to local legend, the ice-topped Mount Kazbeg in Kazbegi, Georgia, in the southern Caucasus, was Prometheus's mythical prison.

The word Promethian is sometimes used today to indicate an especially horrible punishment. The name Prometheus is used in the Ridley Scott film of the same name, where Prometheus is the name of the spacecraft that sustains human life in alien space territory.

Prometheus appears in the Periodic Table as promethium, a radioactive metallic element with the atomic number 61 and the symbol Pm.

Pandora

Pandora and Prometheus are known as shepherd moons. They are both tiny, irregular moons. Pandora's shape is somewhat potato-like. Both shepherd moons suffer from the effects of endless bombardments by fine particles near them.

Fig. 10.7 A crow is the symbol of Hope, at the bottom of Pandora's box. (Illustration by the author)

In mythology, Zeus forced Hephaestus, the smith-god, to create Pandora to avenge both Prometheus's brother, Epimetheus, and especially the whole human race for gaining fire, which he had wanted to belong solely to the gods. Pandora was beautiful, the most beautiful woman ever created. She had to be beautiful or else Epimetheus might have refused to accept her. However, Pandora was also lazy and stupid. Depending on which version of the myth is to be believed, she opened a box or a jar that contained horrible things called the Spites, designed to make the lives of humans considerably less enjoyable. These Spites included Sickness, Insanity and Old Age. After releasing these Spites upon humankind, the only thing that remained was Hope, personified by a crow that clung to the rim of the box or jar (Fig. 10.7).

Telesto

Telesto and Calypso are both tiny moons, known as the Tethys Trojans since they co-orbit with the much bigger moon Tethys. These two moons are irregular in shape, almost unblemished apart from the pockmarks of Oceanus and Tethys. Telesto is associated with success.

Calypso

This is the second of the Tethys Trojan moons.

Calypso was a sea nymph or a beautiful goddess, depending on the variation of the myth chosen. She took care of Odysseus, the hero of Homer's epic poem *The Odyssey*. She took such good care of him that together they produced two sons, Nausithoos and Nausinoos. She wanted Odysseus to remain with her forever, and she even offered him the gift of immortality. He refused this, as he wanted to get home to his wife, the lovely Penelope. After 7 years, Zeus ordered Calypso to release Odysseus, which she reluctantly did.

Helene

This is a Trojan moon.

Helen is famous for her beauty. She had 'a face which launched a 1,000 ships.' She was the daughter of Zeus, who had disguised himself as a graceful, snow white swan, and Leda. She supposedly hatched from an egg, and she had snow-white, delicate skin, reflecting her half-avian parentage. Helen, before eloping to Troy, was a priestess of the Spartan Moon goddess, so she might have had her face painted with lead white on ritual occasions.

Helen's beauty was both a blessing and a curse. When Helen was a young girl dancing near the river, she was raped by the old man Theseus, who was stunned by her beauty and felt compelled to have sex with her. Helen's first husband was Menelaus, king of Sparta, so she was Helen of Sparta before she was Helen of Troy, and she eventually was taken back to Sparta by Menelaus.

The whole myth about Paris and Helen was pre-destined, so it could therefore be seen as somewhat unfair that she has been blamed for the Trojan War. Tyndareus, the man whom Helen knew as her father (her real father being Zeus), made sacrifices to the gods. However, he forgot to make a sacrifice to Aphrodite. Aphrodite was furious and decided to avenge herself by making Tyndareus' daughters infamous for cheating on their husbands.

It was also prophesied that Paris would be the destruction of Troy. This led to Priam, ruler of Troy, asking a servant named Agelaus to kill the baby Paris. He couldn't do it and tried to dump the innocent baby on Mount Ida. A bear allowed the baby to suckle from her. Agelaus felt very guilty and returned to Mount Ida a few days later. He was relieved and delighted to see the child safe and well, so he raised the child as his own. As a young man, his beauty was equivalent to that of Helen. This physical attractiveness plus his cleverness betrayed his true, noble birth.

The gods were pleased with Paris's amusing antics when under Agelaus's care, and Zeus chose him to judge who was the most beautiful of the goddesses, Aphrodite, Hera or Athena. This was a somewhat poisoned chalice, as the goddesses who weren't chosen were guaranteed to be angry with him. All three goddesses offered him enticing bribes. Aphrodite promised him the most beautiful woman in the world as his wife. She gave him a golden apple, the symbol of her promise to him. Unsurprisingly, Paris chose Aphrodite to win the most beautiful goddess title. The other two goddesses went off to plot the demise of the famous city of Troy, as Paris was now a Trojan prince, having been reunited with his true father, Priam.

Paris was the guest of King Menelaus of Sparta when he first set eyes on the radiant Helen. He is supposed to have written the words 'I love you' in blood-red wine on the table while staying there, abusing the hospitality of Menelaus, Helen's husband. Helen was attracted to Paris and left with him of her own free will, taking with her gold and valuable objects from Menelaus. Helen and Paris were eventually married in Troy. All of Troy fell in love with her and refused to return her to Menelaus. Helen, being a daughter of Zeus, had special powers. She found stones in Troy that could produce blood when rubbed together. She recognized it as an aphrodisiac. As a sorceress, Helen could easily have used this, not only to control Paris but the whole of Troy. The Trojan War was then inevitable. It lasted for 10 long years until it was captured with the help of the famous wooden horse that contained Spartan fighters.

Helen could be compassionate. She drugged wine to eliminate bad memories and made Telemachus, the son of Odysseus, an outfit for his bride.

Helen had five husbands: Theseus, Menelaus, Paris, Deiphobus and finally Achilles. Paris died from archery wounds. Deiphobos forced Helen into marriage after she was caught trying to escape from Troy. Deiphobos was killed, literally stabbed in the back, probably by Helen herself.

Famous warrior hero Achilles first had sex with Helen in a dream sent from his mother, Thetis. He enjoyed it so much that he fell in love with her. He died shortly afterwards, through no fault of Helen's.

Menelaus had resolved to murder Helen. But apparently, she took off her clothes in front of him and he changed his mind and took her back with him to Sparta as his queen.

The Irish poet William Butler Yeats was inspired to write a poem about the infamous Helen, who ended up betraying Troy:

"No Second Troy
Why should I blame her that she filled my days
With misery, or that she would of late
Have taught to ignorant men of most violent ways,
Or hurled the little streets upon the great,
Had they but courage equal to desire?
What could have made her peaceful with a mind
That nobleness made simple as a fire,
With beauty like a tightened bow, a kind
That is not natural in an age like this,
Being high and solitary and most stern?
Why, what could she have done, being what she is?
Was there another Troy for her to burn?"

Helen proved both her loyalty to the Greeks and her treachery towards the Trojans by not betraying the disguised Odysseus when he entered Troy the first time. The second time he entered Troy with Diomedes, both in disguise, Helen assisted them and arranged to deliver Troy to the Greeks by shining a light from the top of the fortress. This was the signal arranged with Odysseus for the return of the Greek fleet. She also hid all the weapons from Deiphobus' dwelling just to make sure that he would not be able to fight back.

The Greeks saw that Helen was not even injured during the invasion and initially wanted to stone her to death, but when they saw her beauty, they put down their stones. She was the most beautiful woman on Earth, after all. King Priam of Troy and his son, Hector, did not blame Helen for causing the Trojan War, as they knew that it was the gods that had done that and that they favored Helen. During the decade of the Trojan War, Helen grew bored of Paris and even felt contempt for him. She envied the genuine love that she saw that Hector and his wife Andromache shared until Hector's slaughter by Achilles.

Helen had two children by Menelaus—a daughter Hermione and a son, Nicostratus. When she ran off with Paris, she took Nicostratus with her but not Hermione. Helen also had a daughter named Helena by Paris. Helena is believed to have been killed by Hecuba, mother of doomed prophetess Cassandra, by King Priam.

The sight of Helen when she walked on the walls of Troy raised the anger of both the Greeks and the Trojans. They blamed her for all the loss of life in the war but were also entranced by her beauty. Not only did Helen have aphrodisiac stones at her disposal as a sorceress, she had a herb known as helenium, which protected her against poisonous snakes.

Helen's fate is the subject of several myths. The most well-known is that Helen and Menelaus returned to Sparta together with Menelaus, living in the Elysian Fields when he died with his immortal bride, Helen. A legend from Rhodes tells a very different story. They believed that when Menelaus died, his two sons, Nicostratus and Megapenthes, condemned Helen and banished her from Greece. She traveled to Rhodes with Polyxo.

Polyxo was not exactly a friend of Helen's, as her husband had been killed in the Trojan War. Polyxo had a score to settle, as she blamed Helen for the death of her husband. She disguised some servants as the Erinyes or Furies, whose task was to frighten Helen to death. They succeeded, and Helen hung herself.

Another myth tells of Iphigenia, daughter of Agamemnon and Clytemnestra, offering up Helen up as a sacrifice. There is also another story where Thetis, mother of Achilles, killed Helen because she was angry at the death of her son by Paris shooting him in the heel, his only weak point and where the phrase 'Achilles heel' comes from. These myths involving Helen's death were probably told because there was a need for people to believe that Helen should not escape unharmed and that her actions should have had serious consequences.

The moon Helene shares its orbit with the small moon Polydeuces and the much larger Dione. Helene is irregular in shape and travels slightly ahead of Dione.

Pan

Pan was Zeus's foster brother. Pan's father was Hermes/Mercury disguised as a ram or goat.

Fig. 10.8 Demigod Pan. (Illustration by the author)

He and the nymph Echo were the parents of Lynx. Pan was half man and half goat. He had goat-like horns, legs, a tail and a beard, but otherwise resembled a man (Fig. 10.8). He lived in Arcadia, guarding sheep, cows and of course, goats. He had the gift of prophesy, which Apollo persuaded Pan to give to him.

Daphnis

The name means "laurel." Daphnis was the beautiful young son of Hermes, who had been taught by Pan soulful music on his pipes. A jealous, possessive nymph called Nomia made poor Daphnis swear on his eyesight that he would never cheat on her. However, another nymph called Chimaera got Daphnis drunk and had sex with him. Nomia was furious and blinded him. He died soon afterwards, and his father, Hermes, transformed him into a phallic-looking pillar.

Aegaeon

In Homer, Briareus was also known as Aegaeon. He was one of the 100-handed giants who attacked the Olympian gods.

Methone

Methone was a city in Philoctetes. She was one of the daughters of Alkyoneus, a giant.

Anthe

This was another of the giant Alkyoneus's seven beautiful daughters.

Pallene

This is yet another of Alkyoneus's daughters, all known as the Alkyonides.

Polydeuces

Polydeuces is a small, irregularly shaped moon that moves slightly behind the much larger moon Dione. Its orbit is also shared with fellow Trojan moon, Helene.

Polydeuces/Pollox was the son of Zeus and Leda. Leda had four children, the semi-divine Helen and Polydeuces and the mortal children (sired by Tyndares), Castor and Clytemnestra. At the request of Polydeuces, he and Castor shared the divinity between them, spending half the year with the dead in Hades and the remaining 6 months on Mount Olympus, in the company of the gods. Together they became the constellation Gemini. Roman soldiers believed that their presence on the battlefield was a sign of victory.

Both Castor and Polydeuces were among the Argonauts who sailed on the ship, the *Argo,* with Jason, Heracles and Orpheus on their quest to find the mythical Golden Fleece.

There are five members of the Inuit group of small, irregularly shaped moons: Kiviuq, Ijiraq, Paaliaq, Siarnaq and Tarqeq.

Kiviuq/Quiviuq/Kiviuk/Qooqa

Kiviuq is an irregular moon discovered in 2000. Kiviuq was a shaman and an important hero in Arctic Inuit mythology. He had countless adventures and is believed to still be alive today, although he is apparently gradually becoming a figure of stone. His heart remains very much alive—for the moment—but when it

Fig. 10.9 In the myth of Kiviuq, a snow bunting led the way to land. (Illustration by the author)

stops, this will be fatal for the whole world, not just for Kiviuq. There are many stories about him. Michael Kusugak tells the first story of Kiviuq in the book *The Curse of the Shamen*, where our hero found himself out to sea while still a child. He'd just witnessed other children sink to the bottom of the sea when it became rough. However, Kiviuq was relentless and remained paddling. Eventually he fell asleep, waking up in the middle of the ocean with no idea where land was. Then, a snow bunting appeared on his canoe and flew in front of Kiviuq, showing him where land was (Fig. 10.9). He eventually arrived at this unfamiliar territory.

There is also a story, The Fox Wife, involving Kiviuq as an adult. Kiviuq put up his tent and went off hunting for the day. He came back at night time and found a meal cooked for him. After a few nights, he wanted to know who was providing these wonderful meals. He left pretending to go hunting and lay in wait for this mysterious person. Eventually, a white fox entered his tent. Because she believed that she was alone, she removed her fox-skin and started her cooking.

Kiviuq snatched her fox-skin and refused to return it until she promised to marry him. The fox-woman agreed because she wanted her fox-skin back. (This story goes on and on.)

Ijiraq

Ijiraq is a moon of Saturn discovered in 2000.

In Inuit mythology, Ijiraq was a terrifying evil spirit or bogeyman figure, who took children away from their parents, dumping them somewhere remote and unfamiliar.

Paaliaq

Another irregular moon discovered in 2000, Kiviuq, Ijiraq and Paaliaq all appear light red in color. It is conjectured that the Inuit moons were created when a larger moon or celestial object fragmented.

Fig. 10.10 An owl lifted the curse of Paaliaq. (Illustration by the author)

Paaliaq is an Inuit shaman character, the title character in Michael Kusugak's *The Curse of the Shaman*. Paaliaq, in a bad-tempered moment, cursed a baby called Wolverine and then tried to revoke this curse later, without success. This was because his magical animal, a squirrel, covered its ears when Paaliaq tried to revoke his curse. This caused poor Wolverine severe problems when he became an adult. The curse was eventually lifted when the shaman, Paaliaq's magical squirrel, was eaten by an owl (Fig. 10.10).

Siarnaq

This is the biggest of the Inuit moons, again light red and discovered in 2000. Siarnaq is a goddess of animals in mythology.

Tarqeq/Tarquiup

This was discovered later, in April 2007. The moon is named after the Inuit Moon spirit, Tarqeq, who is a symbol of fertility and animal spirits. When hunters fail to locate and catch food, and people went hungry and bad weather occurred, it was believed to be the punishment of Tarqeq.

Skathi/Skadi/Skade

Skadi was a frost giantess in Norse mythology. She was the wife of Njord, a Vanir god. Skadi had lived in her father's home of Thrymheim, hunting in the icy mountains, in contrast to Njord, who loved the warmth of Asgard, the realm of the gods where the two had first met. They compromised by spending three nights in Thrymheim followed by three nights in Asgard. In a variation of this myth, the gods of Asgard murdered Thiassi, Skadi's father. Skadi demanded a husband as compensation. They agreed on the condition that she chose her husband purely by seeing the feet, not the rest of the body or the face. She chose Njord, a sea god. Being a spirit of winter, they were incompatible, as she loved the bitter cold, snow and ice more than Njord's domain of cozy, Sun-heated bays. Njord's warm passion could only keep the chilly Skadi for 3 months of each year. Skadi was a goddess of winter, including such winter pursuits as hunting and skiing. Skadi was the mother of Freiyja, goddess of love and fertility, and Freyr, god of the Sun and the rain.

Albiorix/Toutatis

This was the god of tribal connection and harmony in Celtic mythology. Associated with Mars as Mars Albiorix. Albiorix was the king of the world. It is most likely that Albiorix was the father or chief deity of the Albici tribe of southern Gaul, whose name was adopted by this tribe. He was believed to have been a mountain god in Gaul, but the association with the Roman god Mars probably comes from his defensive and war-like character.

Bebhionn

This giantess was famed for her enticing, seductive beauty. She was from the west coast of Ireland in Celtic mythology.

Erriapus/Erriapo

This is a giant from Gallic mythology.

Skoll

Skoll was the son of a giantess living in Iron Wood. He was pre-destined, according to Norse mythology, to capture the Sun during her travels across the sky. He would then swallow the Sun, who had just given birth to a daughter, a new Sun that would rise from the sea at Ragnarok.

Greip

Greip was the daughter of the frost giant Geirrod and sister of Gjalp. Greip and Gjalp tried to murder Thor, the Norse god of thunder and lightning. Their intention was to lift up his chair and then to smash his head on the ceiling until his brains fell out. However, with assistance from an amiable giantess Grid and her servants, Thor was able to push down the chair onto Greip and Gjalp, crushing them to death.

Hyrrokkin

Hyrrokkin was an imposing, formidable giantess. She rode a wolf with poisonous serpents for reins. Norse frost giants (such as Hyrrokkin) hated the gods, as the gods had murdered almost all of the giants' ancestors. However the gods needed Hyrrokkin's great strength to push the boat *Ringhorn* into the water for the funeral ceremony of Balder. Balder had been Odin's favorite son and had died tragically. Odin was the leader of the Norse gods. However, the gods did not trust Hyrrokkin. Odin placed four guards around the wolf, who had been knocked unconscious, so it would not cause problems at the solemn funeral ceremony. Hyrrokkin did what she had been asked to do. She pushed *Ringhorn* into the water, but did it with such force that it crashed. Thor, the thunder god, was furious at her lack of respect for Balder, but was restrained by Odin.

Jarnsaxa

Jarnsaxa was a mistress of Thor and a frost giantess in Norse mythology. Jarnsaxa and Thor, the Norse thunder god, had a son called Magni, who was a giant. At the age of three, he was strong enough to move the heavy legs of the dead frost giant Hrungnir, which had trapped Thor, pinning him down. Thor was so pleased with his son that he gave him Hrungnir's magnificent stallion, the amazing Golden Mane. (This annoyed the chief god, Odin, who believed that he should have been given such a special horse.)

Tarvos

In Gallic mythology, Tarvos was the bull god.

Mundilfari

Mundilfari was a man who dared to offend the mighty Odin, the leader of the Norse gods. When Odin and his brothers, Vili and Ve, carved out the world from the body of Ymir, the original frost giant, they illuminated it by using bits of the Sun, Moon and stars. Mundilfari lived in Midgard, where humans lived. He had a son and daughter who were so amazingly attractive that he named one Moon and the other Sun. This act of hubris and arrogance infuriated the gods, and Odin stole Mundilfari's children and turned them into constellations to guide the real Sun and Moon on their passage across the firmament.

Bergelmir

After creating the Cosmos and Earth, Odin, Vili and Ve gave the only remaining giants a place to live called Utgard. These giants were all the offspring of Bergelmir and his wife.

Suttungr/Suttung

Suttung was a frost giant in Germanic mythology. Two wicked dwarves called Fjalar and Galar killed wise man Kvasir to acquire his magic powers. They used the blood of the dead Kvasir to make a magical wine that provided the drinker with wisdom. Fjalar and Galar had murdered his parents, so Suttung took the wine from them in retaliation. Suttung lacked discretion and spoke rather too loudly and too often about this magical beverage so that the gods heard about it. Odin transformed himself into the evil Bolverk, persuading the frost giant Baugi to tunnel through a mountain where Suttung had hidden the magical drink, leaving his daughter, Gunnlod, in charge. When Baugi had made a tunnel, Odin changed from the evil Bolverk into a serpent. He reached the secret place of the magic wine and changed himself yet again, this time into an attractive one-eyed giant who for 3 days and nights became Gunnlod's lover. Gunnlod was so in love with the handsome giant that she allowed him to drink all of the magical wine. Then Odin metamorphosed into an eagle and flew back to Asgard (Fig. 10.11).

At Asgard, Odin spat out the wine into empty jars. Suttung was furious and tried to pursue him, also in the form of an eagle, but was not quick enough to catch the wily Odin.

Narvi

Narvi was the son of Loki, a giant trickster god, and Sigyn, and he was also the brother of Vali. Odin transformed Vali into a wolf, who attacked his unfortunate

Fig. 10.11 The Germanic and Norse god Odin transformed into an eagle in myth. (Illustration by the author)

brother, Narvi, and killed him. Narvi's intestine, in one myth, were used to chain up Loki underneath the mouth of a venomous serpent.

Hati

In Norse mythology Hati was a wolf, brother of another wolf, Skoll, and son of a giantess who lived in Iron Wood. Hati chased the Moon across the sky.

Farbauti/Faubauti

Faubauti was a giant from Norse or Germanic myth. He and his wife, Laufey, produced a son, Loki the trickster and fire god, when Faubauti hit Laufey with a bolt of lightning.

Thrymr

Thrymr was the chief of the giants. He dared to steal Thor's hammer, named Mjolnir, created by dwarves. As if that wasn't bad enough, he also attempted to force the beautiful Freya, goddess of love, fertility, magic and war into marrying him with the threat of keeping Mjolnir. She refused to marry Thrymr, but the gods told him that she had said yes. The wedding was arranged, but Thor dressed as a bride in Freya's place. Thrymr, despite sensing that something was wrong, brought Mjolnir and presented it to his bride as his wedding present. Thor then flung off his bridal attire, leapt into battle and defeated the giants.

Aegir/Eagor

Aegir was a Germanic sea god. With his wife Ran, he fathered nine daughters, the Waves. He was a god to be feared, only rising to the sea's surface when he chose to destroy ships and kill those who were aboard those ships. He was personified as an old, white-haired man. Vikings would sacrifice prisoners to Aegir before setting off for home after attacking other lands. Aegir's wife, Ran, had a drowning-net. This was taken by Loki, the trickster fire god, to help him to catch the richest dwarf, Andvari, who was disguised as a fish.

Bestla

Bestla was the giant Bolthorn's daughter. She became the wife of Bor, the son of the first man. They had three sons, the gods Odin, Vili and Ve. These sons ended up murdering Ymir, the frost giant.

Fenrir

Fenrir's parents were the Norse fire god Loki and the frost giantess, Angrboda. He was a wolf, predestined to be the doom of the gods (Fig. 10.12).

To try to prevent the prophecy from coming true, Fenrir was taken by the gods to Asgard. He was fierce and fearsome. The Norns, the goddesses of destiny, had warned Odin that he was fated to become Fenrir's victim, so he decided that Fenrir had to be restrained. There was no chain strong enough to hold such a ferocious creature, so the dwarves had to make a magic, ribbon-like rope that was named Gleipnir. Fenrir refused to have Gleipnir around his neck unless one of the gods would dare to put his hand between his ferocious jaws as proof that Gleipnir was harmless. The war god, Tyr, did this, and Fenrir bit his hand clean off, while discovering that Gleipnir was unbreakable. Fenrir was then secured to a rock. His mouth was kept permanently jammed open by a sword, so he couldn't bite anyone. At Ragnarok, Fenrir was freed and, as prophesied, swallowed Odin. Wolves appear in other mythologies, such as in the fairytale of Little Red Riding Hood and J. K. Rowling's terrifying wolf Fenrir Greyback in the Harry Potter novels, which perhaps draw upon Norse mythology.

Fig. 10.12 The god Fenrir had the shape of a wolf. (Illustration by the author)

Surtur/Surt/Surtr

Surtur was a pyromaniac giant. According to Norse mythology, he took pleasure in setting fire to everything at Ragnarok.

Kari

Kari was a god of the wind and son of the giant Fornjot in Norse mythology.

Ymir

Ymir was a Norse frost giant who was born of ice in the emptiness known as ginnungagap. The rest of the frost giants originated from Ymir's armpit sweat. Ymir, being the first live creature, fed on the milk of Audhumla, the primeval cow. This cow licked Buri, the first man free from the ice. Ymir became too powerful, and so was murdered by the grandsons of Buri, the trio of brothers, Odin, Vili and Ve. These brothers drowned all the other frost giants in the blood of Ymir, with the exception of Bergelmir and his wife, who escaped. Ymir's flesh then was made into Earth while his skull became the sky.

Loge

This was the god of fire, son of Gornjot, in Norse mythology. Loge is the name that the composer Richard Wagner gave representing Loki in his opera *Ring of the Nibelung*. This is a variation of an existing Norse myth. Loge tricks Alberich, a rich dwarf, into giving up his extensive hoard of golden treasure, including the ring of Nibelung. However, Alberich cursed this gold, so those who came into contact with it ended up losing their lives.

Fornjot/Fornjotr

This was a giant and the father of Kari and Loge.

The rest of Saturn's moons are awaiting a name at the time of this writing. However, Saturn has two hypothetical moons, supposedly orbiting between the moons Titan and Hyperion. Both moons were claimed as discovered by astronomers, but they have not been seen since their original sightings. They are:

Chiron

Seen first by Hermann Goldschmidt in 1861. Chiron was a wise centaur, a healer and a teacher of Achilles and Asclepius. Chiron's parents were one of Oceanus's daughters, Philyra, and Cronus/Saturn in the form of a horse. When Chiron died, Zeus transformed him into the constellation Centaurus.

Themis

Observed by William Pickering in 1905. Themis was a Titaness and was associated with justice and law. Themis and Zeus were the parents of the Horae, or the Hours.

Chapter 11

Uranus and Neptune in the Icy Depths

The Ice Giant Uranus

The planet Uranus was discovered in 1781 by the astronomer William Herschel, who also found two of its moons, Titania and Oberon. It is an ice giant and appears a shade of bluish-green, perhaps azure or natural turquoise in color (Fig. 11.1). It spins on its axis backwards, and the theory is that something large, perhaps another planet, hit Uranus, causing it to be tilted on its side, so it spins on its side when orbiting the Sun. All of its moons line up with this tilted axis.

A day on Uranus lasts 17 h and 17 min. The planet is four times wider than Earth, with a diameter of 51,118 km (31,763 miles) and takes 84 years to orbit the Sun.

This planet has been said by many to be really boring, but it has of late become more active, with white spots on it. Uranus in older textbooks is sometimes referred to as Georgium Sidus, as this was Herschel's choice of name (after King George III). The planet was eventually given the name of Uranus, the father of Saturn. It has 29 known moons, and it looks as if another moon was obliterated, with the only clues to its existence being thin rings encompassing the planet, which are very dark compared to those of Saturn. Very recently, aurorae were discovered on Uranus. Uranus is represented in the Periodic Table as uranium, with the atomic number of 92 and the symbol 'U.'

© Springer Science+Business Media New York 2015
R. Alexander, *Myths, Symbols and Legends of Solar System Bodies*, The Patrick Moore Practical Astronomy Series, DOI 10.1007/978-1-4614-7067-0_11

Fig. 11.1 A view of Uranus. (Image courtesy of NASA)

Ouranus/Uranus in Mythology

In ancient Greco-Roman mythology, Ouranus/Uranus was the son of Gaia (Earth) and the father of Cronos/Saturn. Uranus was the sky god, father of all the gods known as the Titans, and the god of astronomy. (Ouranos/Uranus means "heaven.")

One of the more well-known and gruesome of the Greek myths involves the son of Uranus, Cronos, who castrated his father with a sickle. Cronos was then attacked by his son, Zeus, who hurled lightning at him in retaliation. Uranus was overthrown by Zeus and seemed to vanish from the sphere or pantheon of the gods.

It is imagined that Gaia, Uranus's wife and mother, was the brains behind the acts of violence. The sliced-off penis of Uranus was not wasted, as some believe that the goddess Aphrodite/Venus was born from it, and out of the imagined vast quantity of blood lost in the castration came the tree-spirits, including the Dryads and the Furies/Erinyes. The furies are, predictably, formidable, nightmarish creatures with their fire and whips and serpentine hair, but the tree-spirits are more benign. In Hesiod's work *Theogony*, Ouranos was a sky god and wouldn't separate himself from Earth, or Ge. This separation was achieved by Cronos's famous castration of his own father, which enabled birth to take place. This led to Aphrodite/

Venus being born from the sea, Night begetting Day, and death, pain, strife and dreams. Most of this mythology originates from Mesopotamian or other pre-Greek tales.

The connection with astronomy springs not only from the fact that Uranus was a sky god. There is also a narrative in which Uranus was a mortal king who instructed the people in the mysteries of astronomy. It is also no accident that the Muse or inspirational goddess of astronomy is named Urania.

The Moons of Uranus

Twenty-four of Uranus's moons are named after characters, including some very minor ones that are featured in the plays of William Shakespeare, while three have names from "The Rape of the Lock," a poem by the English writer Alexander Pope. The rest are unnamed at the time of this writing.

The first five moons are much bigger than the others.

Titania

Titania is the biggest of Uranus's moons, at 1,578 km (986 miles) in diameter. Discovered by William Herschel, it is a frozen world. In Shakespeare's play *A Midsummer Night's Dream*, Titania is the queen of the fairies. Her husband, Oberon, had Puck put love-juice from an enchanted flower on her eyelids. This causes her to fall in love with the first creature she sees when she awakens. This is Nick Bottom, a weaver who happens to be wearing the head of a donkey or ass, and therefore looks ridiculous. In the end, she believes that this incident was nothing but a dream and is reconciled with her husband.

Oberon

Oberon was also discovered by William Herschel and is 1,523 km (1,974 miles) in diameter. It has many craters on its surface. In Shakespeare's *A Midsummer Night's Dream,* Oberon is king of the fairies. He becomes annoyed with his wife, Titania, as she stole a boy from an Indian king to whom she gives all her attention, The jealous Oberon puts love-juice on his wife's eyelids when she falls asleep. This makes her fall in love with the insane-looking Nick Bottom, who is wearing a donkey's or ass's head (Fig. 11.2). Eventually Oberon rescues the boy and reverses the love-spell. Oberon and Titania dance together and are reconciled.

Fig. 11.2 Titania was enchanted at the behest of her fairy husband Oberon. (Illustration by the author)

Umbriel

Umbriel is a heavily cratered moon, 1,170 km (727 miles) in diameter. In Pope's "The Rape of the Lock," Umbriel is an Earth demon and mischievous sprite that has black wings. He can make spots appear on a beautiful face and can plant nagging doubts in the minds of jealous husbands.

Ariel

There are a few craters on Ariel, but the moon has many deep gouges in its surface. Ariel is now a world without signs of activity, but it is conjectured that it used to have been more active. It is 1,160 km (720 miles) in diameter. In Pope's "The Rape of the Lock," Ariel is a watchful sprite, a spirit of the air that can, mischievously, cause small cracks in vases. Ariel also appears in Shakespeare's play *The Tempest*. He is a hardworking sprite who can change shape. Though he is Prospero's slave, he wins his freedom by the end of the play. Ariel can fly, turn into things and become invisible. Ariel, on Prospero's orders, created the storm that caused the shipwreck at the beginning of the play. Prospero had rescued Ariel, who was Sycorax, the witch's servant. She had trapped him in a split pine tree for 12 long years before Prospero came along.

Miranda

Miranda is the nearest of the biggest moons to the planet Uranus itself. It has a diameter of 474 km (295 miles). Miranda looks as if it's been bashed apart and then the bits stuck back together. Miranda is the fifth largest of Uranus' moons. It appears to have rather unique features on its surface such as markings and grooves. Miranda is Prospero's daughter in Shakespeare's play *The Tempest*. She falls on love with only the third man that she has ever seen, Ferdinand, son of the king of Naples. They eventually become betrothed to each other.

Puck

In Shakespeare's play *A Midsummer Night's Dream* Puck, or Robin Goodfellow, is described as Oberon's jester and servant, who finds the herbs that are used to create the love potions. However, he sometimes makes mistakes, and the wrong humans end up taking it. This results in the four human lovers quarrelling, but eventually Puck and Oberon get it right and the story ends with two happy human couples, Lysander and Hermia and Demetrius and Helena.

Cordelia

This is the youngest daughter of King Lear in the Shakespeare play of the same name. Unlike her two sisters, Goneril and Regan, Cordelia is unmarried, and also unlike her sisters, she refuses to flatter her father. The duke of Burgundy refuses to marry her when King Lear takes away her dowry, but the king of France still wants her, recognizing her true goodness, so she leaves with him. She is reunited with her father much later but is distressed when she discovers that he is mad. She ends up being hung despite being innocent of any wrongdoing. Lear recovers his sanity too late and lamented her loss while holding her dead body in his arms.

Ophelia

In Shakespeare's famous tragedy, *Hamlet,* Ophelia is the daughter of Polonius and sister of Laertes. She is in love with the title character, Hamlet, prince of Denmark, despite her brother telling her that Hamlet cannot choose his own wife. On her father's advice, she repels Hamlet's letters and will not let him make love to her, even though she wants him to. Hamlet tells her that he loved her once but no longer feels it. She sees how Hamlet has changed since the suspected murder of his father. Hamlet kills Polonius (who is hiding behind the arras or tapestry). Ophelia goes mad and drowns. It is unclear whether it was suicide.

Bianca

Bianca is the beautiful, even-tempered sister of Katherina, the shrew of the title of Shakespeare's play *The Taming of the Shrew*. Unfortunately, Katherina, being the older sister, had to be married first. Eventually, Katharina is "tamed" by Petruchio and then Bianca can marry Lucentio, whom she loves.

Cressida

Cressida is a Trojan in Shakespeare's play *Troilus and Cressida*. Troilus, son of the king of Troy, Priam, is in love with her. She plays hard to get but eventually admits that she loves Troilus. They promise to be true to each other. However, Cressida's father, Calchus, who had defected to the side of the Greeks, wants his daughter to be swapped for a captured Trojan leader. Cressida doesn't want to leave Troy and the Trojans. Troilus and Cressida exchange tokens. Troilus gives Cressida a sleeve. Troilus later watches Cressida with the Greek Diomedes from a distance. He thinks she is flirting and sees her give Diomedes the sleeve that Troilus gave to her. She does snatch it back and refuses to tell Diomedes whose sleeve it is. She also writes Troilus a letter, but he dismisses it as mere words and tears it up. This ends their love affair.

Desdemona

Desdemona is the beautiful wife of the title character of Shakespeare's play *Othello*. Othello, a Moorish general, and Desdemona are in love. But Othello's ensign, named Iago, hates Othello and tries to make him jealous by claiming that Desdemona and Cassio, Othello's lieutenant, are lovers. Othello believes the evil Iago and doubts the innocence of his wife. Desdemona is unfortunate enough to drop her handkerchief, which was her first gift from Othello and was embroidered with strawberries. Emilia, Iago's wife, picks up this handkerchief to return it but has it snatched from her by her husband. Iago plants the handkerchief in Cassio's bedroom so that Othello will find it there and tells Othello that he heard Cassio talking of Desdemonia in his sleep and that he saw Cassio wiping his beard with the strawberry handkerchief. Consequently, Othello demands that Desdemona show him the handkerchief, which she is unable to do. Othello asks Emilia if Cassio and Desdemona have ever been alone together. Emilia insists that she hasn't and protests Desdemona's innocence. Desdemona also protests her innocence to Othello, but to no avail. Othello murders her. Upon learning of his wife's innocence, Othello kills himself.

Juliet

This is a minor character in Shakespeare's play *Measure for Measure.* She is unmarried and heavily pregnant by Claudio. They love each other, but Claudio is condemned to death for impregnating her. Claudio is eventually pardoned by the duke, Vincentio.

Juliet is also the very young lover of Romeo in Shakespeare's play *Romeo and Juliet.* Their families are enemies, so their love is forbidden. Romeo is banished from Verona after a fight ends in his murdering Tybalt (a cousin from Juliet's Capulet family), and Juliet is to marry Paris. However, Friar Laurence gives her a drug that stops her pulse and makes her seem dead. Romeo thinks she is dead and takes poison and dies. Juliet wakes up and sees that Romeo is dead and so stabs herself and also dies.

Portia

This is the rich, intelligent heiress in Shakespeare's play *The Merchant of Venice.* She makes her suitors choose to open either a gold, silver or lead casket. The prince of Morocco chooses the gold casket, and the prince of Aragon the silver one. Both are turned down. Bassanio chooses the lead casket and is accepted. Bassanio tells Portia about his friend Antonio, who is unable to pay back Shylock, a Jewish moneylender, but Shylock is demanding a pound of flesh in payment. In court Portia, dressed as a man, says that if Shylock sheds even a single drop of blood when taking his pound of flesh from Antonio, his land and goods would be forfeited. Shylock chooses not to take his pound of flesh and Antonio is set free.

Rosalind

Rosalind was the daughter of the old duke in Shakespeare's play *As You Like It.* Her father has been banished but she remains at court. Rosalind and cousin Celia watch a wrestling match between Charles, the court wrestler, and Orlando. Orlando wins.

Orlando and Rosalind fall in love. Rosalind is then banished and Celia decides to accompany her. Rosalind dresses as a man, Ganymede, and Celia disguises herself as Aliena. They go to the forest to find Rosalind's father and eventually all meet up, and Rosalind and Orlando are married.

Cupid

The Roman god of love, Cupid is the Greek equivalent of Eros. He is NOT a character in Shakespeare's plays but is often referred to in them. Sometimes he shoots golden-tipped arrows.

Belinda

Belinda is the beautiful woman in Alexander Pope's poem "The Rape of the Lock." It is her lock of hair that is referred to in the poem's title. The story of the poem is based on an actual occurrence. A man called Lord Petre was so bewitched by Arabella Fermoy that he took a lock of her hair without her permission!

Perdita

Perdita is the daughter of Leontes, the king in Shakespeare's play *The Winter's Tale*. Because Leontes became convinced that his wife, Hermione, had been unfaithful and that baby Perdita was the product of this, they are estranged for 16 years. Perdita is separated from her mother and brought up by an old shepherd and his wife. Florizel, a prince of Bohemia, falls in love with Perdita. Eventually Perdita and Leontes are reunited. Paulina, an old loyal servant, casts a spell on a statue of the dead Hermione, which comes to life, acknowledging Perdita as her daughter. The story has echoes of Oedipus about it, with Perdita being lost and found by someone who brings her up. It also echoes the ancient Greek belief in the animated quality of statues.

Mab

Mab is not a character in Shakespeare's plays but is referred to by Mercutio, a friend of Romeo's, in *Romeo and Juliet*. Mab is head of the Irish fairies in mythology. She plants desires in the dreams of men while they are asleep.

Francisco

This is a minor character in two of Shakespeare's plays. In *The Tempest* he is a lord who is shipwrecked with Antonio, Sebastian, Gonzalo and Adrian. He believes that Prince Ferdinand, son of the king, Alonso', is alive, even though Alonso believes him to be dead.

Francisco is also a guard in *Hamlet*. He is on guard at the beginning of the play. Barnardo relieves him from duty at midnight after a shift that is very quiet.

Caliban

This character is based on Prince Calibos from Greek mythology, son of Themis, who was punished by Zeus with deformity and forced to live in the swamps, forbidden to marry the beautiful Andromeda. In Shakespeare's play *The Tempest,* Caliban is the son of the witch Sycorax and a slave to Prospero. Miranda, the beautiful daughter of Prospero, teaches him language, and he lusts after her.

Stephano

This minor character in Shakespeare's play *The Tempest* is a drunken butler who becomes shipwrecked on Prospero's island. He gives Caliban wine. Caliban then decides that he is his new god! Caliban tells Stephano about Prospero and Miranda. Stephano wants to kill Prospero and marry Miranda but is usually too drunk to do anything about it.

Trinculo

This is a jester who is reunited with Stephano on Prospero's island in *The Tempest*. He crawls behind Caliban's cloak, making Stephano believe that Caliban has four legs until he recognizes Trinculo's voice.

Sycorax

Sycorax does not actually appear but is referred to in *The Tempest*. She is a bent-over old hag or witch and the mother of the ugly Caliban.

Margaret

Margaret is a servant to Hero in Shakespeare's play *Much Ado About Nothing*. Margaret dresses in her mistress Hero's clothes and leans out the window of her mistress's chamber or bedroom and calls the roguish Borachio by the name of Claudio. This is part of the complex deception by Borachio, who is promised a large amount of money if he can break up the proposed wedding between Hero and Claudio (the idea being to persuade Don John to tell Claudio and Don Pedro that Hero loves Borachio). Then Claudio would believe that Hero was unfaithful and the marriage would be called off. This almost works, but Borachio is forced to confess his part in the plot. Borachio also says that Margaret was unaware of what was really going on.

Prospero

This is the former duke of Milan in Shakespeare's play *The Tempest*. He was deposed by his own brother, Antonio, and then set adrift in a flimsy boat with neither a mast nor sail with his 3-year-old daughter, Miranda. Another man, Gonzalo, gives them food, water, clothes and books, which are the source of Prospero's power. Prospero has studied many things, including astronomy, and believes in the stars. He wants Miranda to marry Ferdinand, the prince of Naples, and so engineers a meeting, though he forces Ferdinand to work for him before he is willing to let the two marry.

Setebos

Setebos does not actually appear in but is referred to in Shakespeare's play *The Tempest*. He is Caliban's god until Caliban meets Stephano (who gives him wine).

Ferdinand

Ferdinand is Alonso's son in *The Tempest*. Alonso is the king of Naples. Ferdinand is only the third man that the heroine, Miranda, has ever seen. Miranda instantly trusts him. Ferdinand falls in love with her and asks her to marry him. She says 'Yes!'

The Ice Giant Neptune

Neptune is the outermost of the recognized planets in our Solar System now that Pluto's status has been downgraded. It was discovered through mathematics. There is some controversy over exactly who made the discovery. Most credit this achievement to Johann Galle in 1846.

Like Uranus, Neptune is an ice giant and has five dim arcs that encircle it. It appears a deeper shade of blue than Uranus, more like cobalt than the turquoise color of Uranus, because of its abundant methane (Fig. 11.3). Neptune has an enigmatic internal wind speed of up to 1,000 miles/hour. There are many ice-giant

Fig. 11.3 The almost purple planet of Neptune, colored by its high methane content. (Image courtesy of NASA)

planets like Neptune in other solar systems. The Neptune in our Solar System has 13 moons and features in the Periodic Table as neptunium, with the atomic number 93 and the symbol 'Np.'

The Mythology of Neptune/Poseidon

Neptune is named after the sea god from Roman times whose Greek equivalent is Poseidon. It is called Neptune because it appears blue in color. After the overthrowing of the Titans, a dynasty of gods, including Cronus/Saturn, Atlas, Prometheus, Rhea and Hyperion, the next generation that took over were known as the Olympian gods, with Zeus/Jupiter as the supreme leader. When dividing up the kingdom, Poseidon was given the ocean and seas, although he is also known as the Earth shaker, responsible for earthquakes, tsunamis and general watery mayhem. Poseidon was regarded as a violent, tempestuous god of enormous strength. Son of Cronos and Rhea, he was married to Amphirite, a sea nymph.

Like his brothers, Hades/Pluto and Zeus/Jupiter himself, he was not a god that anyone would want to deliberately or even accidently displease! In Homer's *Odyssey*, Poseidon/Neptune makes the hero Odysseus's sea journey treacherous and perilous with ferocious, almost fatal waves because Odysseus and his men plied the Cyclops Polyphemus with wine until he became drunk. They then stuck a hot pointed pole in his single eye. Unfortunately for Odysseus and his men, Polyphemus was a son of Poseidon. Polyphemus prayed to his father, requesting that Odysseus should arrive home late, that his men would all die, and that when he finally arrived back in Ithaca, there would be further problems.

Poseidon heard this prayer and made all of Polyphemus's requests happen. Odysseus's journey took 10 long years. Poseidon becalmed his ship at one point and caused such furious storms that his ship was smashed to pieces. When he eventually returned home to Ithaca, the people there, assuming that he was dead, were trying to persuade his faithful wife, Penelope, to remarry, something that she was very reluctant to do, as she still loved Odysseus.

King Minos of Crete was also foolish enough to anger Poseidon. He was supposed to sacrifice a beautiful, pure white bull to Poseidon, but he couldn't be bothered. The bull was then made to have sex with the King Mino's wife, Pasiphae, which resulted in the birth of the Minotaur, a man with the head of a bull.

Poseidon, like his brothers Zeus and Hades, would be judged a serial rapist by twenty-first century standards. The Gorgon Medusa had not always been the grotesque creature with coiling serpents—poisonous adders—for hair who could turn men to stone by looking at them. She had once been a beauty famed for her incredible head of hair, and many men wished to marry her. Unfortunately for the lovely Medusa, Poseidon saw her and wanted to have sex with her. She had no choice in the matter. Even more unhappily for the radiant Medusa, Poseidon's chosen place for having sex was inside the shrine of Minerva/Athena, virginal goddess of war

Fig. 11.4 The ram with the golden fleece. (Illustration by the author)

and wisdom. Minerva placed her shining breastplate in front of her eyes, truly horrified. Then, rather unfairly, she turned Medusa's beautiful tresses into venomous serpents.

Poseidon was a more powerful god than Minerva was, so Medusa was easier to punish, although she could have enlisted her father Zeus's help. The union of Poseidon and Medusa produced Pegasus, the winged horse. Poseidon also raped Theopane after he had metamorphosed into a ram (Fig. 11.4). This union produced the ram with the Golden Fleece. This was the very fleece that Jason and the Argonauts stole, despite it being guarded by a dragon that never slept. Another rape victim of Poseidon involved a beautiful woman called Caenis. After being raped, she was allowed to request a gift from Poseidon. She said that she didn't want to be raped again or be a woman any longer. Poseidon then transformed her into a man called Caeneus.

The power of Poseidon/Neptune is illustrated in Plato's myth of Atlantis. This mythical island was swallowed up by the sea, as the people who lived on Atlantis had upset the god.

Poseidon lived deep under the sea. His chariot was driven by seahorses. Even his brother Pluto/Hades was afraid that Poseidon's earthquakes would damage his kingdom, which lay beneath that of Poseidon's.

Neptune's Moons

Neptune has 13 named moons, of which Triton is the most intriguing. They are named after lesser sea gods, nymphs and children of Poseidon/Neptune himself.

Triton

Triton is Neptune's largest moon (Fig. 11.5). It is about a third of the size of Earth's own moon and is larger than Pluto. It was discovered in 1846 and is believed to be the coldest celestial body in our Solar System. It is conjectured that Neptune's strong gravitational field attracted and seized Triton. Triton revolves around Neptune in a circular route but spins in the opposite direction to its parent planet. It is believed that Neptune's strong gravity will eventually cause Triton to become fragmented. Triton is a tiny frozen world with hot springs on it.

Triton is sometimes depicted as a merman and sometimes as a man-headed fish. He can be seen holding the three-pronged trident of Neptune or blowing through a conch-shell trumpet that can stir up the power of the waves. Triton is also the river god that raised the goddess Athena.

The other moons of Neptune are almost all minute, spinning in non-circular orbits.

Fig. 11.5 Neptune's largest moon Triton. (Image courtesy of NASA)

Nereid

Nereid has an elliptical orbit. It was discovered almost a 100 years after Triton, in 1949. Nereus was a shape-shifting sea god who married Doris, a goddess of the sea. Together they had 50 daughters, collectively known as the Nereids, nymphs of the water, or mermaids. The Nereids were attendants of Thetis, another sea goddess who was the mother of the hero Achilles. The Nereids enticed Hylas, one of Jason's Argonauts, down into their pool, where he drowned.

Naiad

The moon Naiad was discovered much later, in 1989. The Naiads were water nymphs, usually found near fresh water. They are mostly considered somewhat benign creatures but can be dangerous temptresses at times. They enticed attractive young men into the water, somewhat similar to the Lorelei from Germanic mythology who like to entice young men into the water, where they are drowned. The Naiads were mother to the 50 Danaides, all of whom murdered their husbands on their wedding night.

Thalassa

The moon Thalassa was also discovered in 1989. Thalassa means "of the sea." Therefore, Thalassa was most likely a sea goddess who might have been a daughter of Poseidon or of Aether and Hermera.

Despina/Despoena

Despina was yet another moon discovered in 1989 and was also supposed to be a daughter of Poseidon. She was a nymph. Her conception was unusual in that Poseidon had metamorphosed into a stallion, and her mother, the goddess Demeter, morphed into a mare. This union also produced a brother for Despina, the wild horse Arion.

Galatea

The most likely explanation of Galatea's identity is that she was a nymph who was in love with an attractive young man named Acris. However, a Cyclops, the one-eyed

Fig. 11.6 Pygmalion fell in love with his sculpture of Galatea, who was later turned into a real woman by Aphrodite. (Illustration by the author)

giant Polyphemus, was in love with Galatea. Unsurprisingly, she was not interested. Polyphemus became consumed with jealousy and crushed the unfortunate Acris with a rock. Galatea turned the dead Acris into a river in Sicily.

There is another ancient Greek myth where a character called Galatea is featured. Pygmalion fell in love with the goddess Aphrodite, who refused to sleep with him. Undaunted, Pygmalion crafted an ivory statue of her, laid it in his bed and prayed to Aphrodite to take pity upon him (Fig. 11.6). She did and brought the statue to life. The statue-woman was called Galatea. The ancient Greeks credited statues with human characteristics such as the ability to die and human emotions.

Galatea means "milk-white." The silvery-white metal gallium (found in LEDs) was named for Galatea. It features in the Periodic Table with the atomic number 31 and the symbol 'Ga.'

Larissa

Larissa is one of the more obscure astromythological references. It appears in mythology as a place. Ovid mentions Larissan Coronus, who was a beautiful young girl. Also, Perseus went to Argos to find his grandfather, Acrisius, only to discover that he had already fled to Larissa. (Acrisius fled from his grandson because it was prophesied that he would be killed by him.) However, Perseus followed him to

Larissa and struck his grandfather on the head with a discus (accidentally!) when he competed in the games, which fulfilled the prophecy. Another death in mythology occurred at Larissa. Ixion was the Thessalian king of Larissa. He wanted to avoid paying his father-in-law, Eioneus, his fee for the privilege of marrying his beautiful daughter, Dia. Ixion then killed Eioneus. Larissa, as a place, seems to have had a bad reputation. Larissa is most likely to have been a local nymph, as nymphs were frequently named after the place they were supposed to have inhabited. Larissa is still a place in Greece.

The name Larissa has evolved in legend after the character Larissa in Boris Parternak's novel, *Doctor Zhivago,* who was the title character's lover and muse for his exquisite poetry.

Larissa was discovered in 1989, when *Voyager 2* flew by Neptune.

Proteus

The moon Proteus has a gray, spotty, lumpy surface, rather like a face with a very bad case of acne (Fig. 11.7).

Proteus was supposedly the offspring of Poseidon. He was a sea god with the ability to change his form whenever he chose.

Proteus is also known as the old man of the sea (Fig. 11.8). He was the only sea god who could instruct Menelaus how to break the spell in the form of a storm cast by the goddess Athena and also how to gain more favorable winds. Menelaus and three others disguised themselves by wearing stinking seal skins and then lying on the shore until they were eventually joined by Proteus's flock of hundreds of real seals. Proteus appeared and slept among his flock of seals. Menelaus and the others seized him, despite Proteus changing shape several times from animals such as a reptile to water and a tree. They held onto him, forcing him to help. Menelaus, who was told to visit Egypt to appease the gods. Then the winds became more favorable.

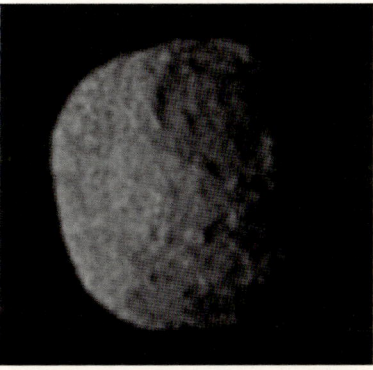

Fig 11.7 A view of Neptune's moon Proteus. (Image courtesy of NASA)

Fig. 11.8 Proteus, or the old man of the sea. (Illustration by the author)

Halimede

Halimede was a Nereid, the daughter of Nereus and Doris.

Psamathe

Psamathe was also a Nereid. Aeacus saw a group of seals swimming under a full Moon. They took off their seal skins and became beautiful, naked young women. Aeacus enjoyed watching them dancing in the moonlight and then decided to steal a seal skin. This belonged to Psamathe. While he owned the seal skin, he controlled its owner, as her power came from the animal skin. He had sex with her, even though she tried to escape from him, and they produced a son, Phocus. After a while, she recaptured her seal skin and escaped from her captor.

Sao, Laomedia and Neso

These are individual Nereids.

Chapter 12

Pluto, a Dark, Distant Underworld

The Ninth Planet, Now a Dwarf

The discovery of the ninth planet did not exactly happen overnight. Its presence had been suspected for a while. In 1846, the British astronomer and mathematician John Couch Adams (1819–1892) and the French mathematician Urbain LeVerrier (1811–1877) were jointly credited suggesting it. Astronomers had noticed that the orbits of both Uranus and Neptune were disturbed. These inconsistencies were interpreted as the influence of a large ninth planet. We now know that this was a mere miscalculation.

The search for a ninth planet was seriously undertaken by two American astronomers, Percival Lowell (1855–1916) and William Pickering (1858–1938). Their faith and sheer enthusiasm, and Lowell's considerable wealth, contributed significantly to Pluto's discovery. It is just a pity that Lowell did not live to see it.

Lowell must have been the envy of most astronomers both then and now, in having the financial means to establish his own observatory in 1894 in Flagstaff, Arizona, which Pickering helped to set up. Initially, Lowell was only interested in the planet Mars, before embarking on a clandestine search for a ninth planet in 1905. When he realized that his quest was proving more difficult than he expected, he dropped the secrecy and publicly referred to his holy grail as 'Planet X.' Pickering joined the search in 1909. They took over 1,000 photographic plates of the suspected area of the Solar System and spent years studying them.

Unfortunately, neither man realized that their elusive holy grail or treasure was right in front of them, as it appeared twice on Lowell's plates. However, both men

© Springer Science+Business Media New York 2015
R. Alexander, *Myths, Symbols and Legends of Solar System Bodies*, The Patrick Moore Practical Astronomy Series, DOI 10.1007/978-1-4614-7067-0_12

were expecting a much larger, heavier celestial object. Lowell believed that Planet X would be seven times the mass of Earth, taking 282 years to orbit the Sun. Pickering believed that this ninth planet would have a mass of less than twice that of our own planet, taking 373.5 years to orbit the Sun. Pluto is actually a mere 0.02 of Earth's mass, smaller than our Moon, and takes 248.6 years to orbit the Sun, so Pickering's calculations were nearer in mass, but Lowell's were not that far out regarding the length of Pluto's orbit of the Sun.

The search for the ninth planet continued after Lowell's death in 1916. (Lowell died in his own observatory!) It was continued by the American astronomer Vesto Slipher (1875–1969). Slipher hired the young amateur astronomer, Clyde Tombaugh (1906–1997). Tombaugh's mission was to find this ever-elusive ninth planet, a somewhat lengthy and detailed, yet exciting, task. Tombaugh's diligence paid off. He photographed the sky like Lowell had and Pickering before him and then compared photographic plates. He was looking for a celestial object that had moved its position slightly when photographed a week earlier. He had a huge technological advantage over Lowell by using a blink comparator—a device that allowed him to look at two photographic plates at the same time. This was far more effective than a magnifying glass.

Clyde Tombaugh is credited with the discovery of the planet Pluto on February 18, 1930, and it was made official on March 13, Lowell's birthday. There was much excitement at the discovery of the first new celestial body in over 80 years, but there was even more public involvement with the new planet's name, as suggestions for a new name flooded in. Constance Lowell, widow of Percival (1863–1954), suggested some names including Constance, Percival and Lowell, but these were turned down, as a classical name was required.

Walt Disney's canine cartoon character Pluto appeared in the same year as the new planet's discovery, linking cartoons with this new celestial body. An English schoolgirl, Venetia Burney, was the one who gave the planet the name of Pluto, even though there were other potential names. These names included Minerva and Cronus. Saturn is the Roman equivalent of Cronus and Minerva is the name of an asteroid, so both these names were already being used. Pluto happened to be the second most popular name with the public. It ticked all the boxes, being a classical name that fitted with its position in the outer Solar System with the added bonus of having the symbol Pl, Percival Lowell's initials.

Venetia Burney was not just an ordinary member of the public. Her uncle, Henry Madan, had named both the moons of the planet Mars. He had the right connections, with one of them sending a telegram to the Lowell Observatory in Flagstaff containing the suggested name.

Pluto's status has been in question since 1936. The scientific conjecture in the 1930s was that Pluto was not a planet at all but a former moon of Neptune that had somehow gotten free. This is no longer believed to be the most likely explanation. In 2004, the asteroid Sedna was discovered, so it was thought that Pluto and Sedna were merely Kuiper Belt bodies as opposed to being legitimate planets. Pluto is now regarded as a planetoid, dwarf planet or asteroid, as there are many such ice 'planets' just like this in our outer Solar System.

Pluto has an elliptical, very tilted orbit and is tinier than Earth's Moon, with a diameter of just 2,274 km, or 1,413 miles. It takes a lengthy 249 years to travel around the Sun and 6 days and 9 h to rotate. Pluto is the only 'planet' that a spacecraft from Earth has not traveled to. A probe is on its way now, however, so we will learn more about this celestial body soon. It is believed that Pluto is blanketed with nitrogen, methane and carbon monoxide. It is extremely cold and icy, and it has been discovered recently that it has an enormous atmosphere.

Pluto features in the Periodic Table as plutonium, with the atomic number of 94 and the symbol 'Pu.' Plutonium is a radioactive metallic element made from neptunium and is used in the nuclear industry.

Pluto's Mythology

Despite Pluto's recent status downgrade, it is as important as the other legitimate planets in our Solar System, due to its intriguing mythology. It is also interesting from the astronomical point of view, as there have been new discoveries.

Pluto and the Greek equivalent, Hades, was a sinister character. He was the brother of Jupiter/Zeus and Neptune/Poseidon and ruled over the Underworld, where souls go after death. He inspired such fear that many were afraid to even utter his actual name and called him 'Lord' or 'the other Zeus.' This might have inspired J. K. Rowling in her Harry Potter novels, as her character Lord Voldemort also inspired such fear that characters in the novels were afraid to speak his name out loud.

Pluto was reputed to ignore prayers and pleading, and, like his brothers, he was not a god whom anyone would want to cross. He liked to allocate especially futile, insanity-inducing punishments. Sisyphus was foolish enough to try to trick Hades. His eternal punishment was to be forced to roll a huge rock up a hill where it always fell back down the hill again. From this myth comes the term Sisyphian, meaning futile and pointless in our English language, an ever-changing animal. The punishment for the Danaides, who were the daughters of the king of Argos, was to pour water constantly into a container where it always dripped out of the bottom. (The Danaides all killed their husbands on the night of their collective wedding.)

One day, Pluto saw the beautiful Proserpina/Persephone, daughter of Ceres and the goddess of spring, sleeping after gathering Lent lilies or daffodils (Fig. 12.1). She had made a garland from the flowers and placed it on her head, where they turned to a beautiful spring yellow color. Pluto decided to steal her from Earth and take her down to his kingdom in the Underworld. Here he made her his queen without the permission of her mother Ceres/Demeter, goddess of the cornfield.

Ceres was devastated when she realized that her precious, innocent daughter had gone missing. She searched high and low for a very long time until she finally came across a clue, an intimate item of Persephone's clothing that she recognized. Ceres realized something must have happened. She was both horrified and furious. She took out her rage on Earth, especially on Sicily, where she found the clothing

Fig. 12.1 Persephone was captured by Pluto while collecting daffodils. (Illustration by the author)

of her daughter. She caused crops to fail and murdered cows and the men who looked after them.

A nymph called Arethusa witnessed this and pleaded with Ceres to spare Sicily. She also told her that she'd seen Persephone in the Underworld. Then Ceres went straight to Zeus/Jupiter, Persephone's father, to ask for her daughter back. He thought that Pluto hadn't done anything wrong, as he had married Persephone and claimed to love her. He did say that Persephone could come back to Earth if she hadn't eaten a single bite of food during her time in the Underworld. Unfortunately for Persephone and Ceres, Persephone had already eaten seven pomegranate seeds (Fig. 12.2).

She could have so easily gotten away with eating the seeds if Ascalaphus had not seen her. He told Pluto what he had seen. In retaliation, Persephone turned Ascalaphus into a screech owl, a symbol of bad luck, ruin and death (Fig. 12.3). Finally, Zeus decided that Persephone should spend half the year with Pluto in the Underworld and the rest of the year on Earth with Ceres.

However, even the usually heartless Pluto was uncharacteristically charmed by the poetic beauty of Orpheus's music in the story of the Orpheus and the Underworld. Orpheus had loved his wife Eurydice, but she had died suddenly from the venom of a serpent and went straight to the Underworld. Orpheus, with the assistance of his beautiful music, managed to persuade Pluto into giving him a

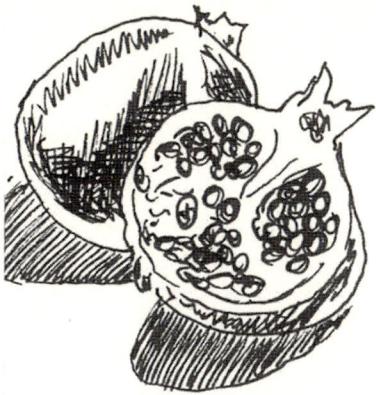

Fig. 12.2 The pomegranate seeds eaten by Persephone doomed her to spend part of the year in Hades. (Illustration by the author)

Fig. 12.3 Ascalaphus was transformed into a screen owl as punishment for his betrayal. (Illustration by the author)

chance to save his beloved wife. He could return to Earth with Eurydice if he did look back at her throughout his return journey. Predictably, he was unable to resist looking back at his beautiful wife, and at that very moment, Eurydice was lost forever to the Underworld.

Hades/Pluto was supposed to have been invisible and so is not represented in art. He wore a helmet that turned him invisible. This concept of invisibility has captured our imaginations since ancient times, from Pluto's helmet to more recent television, film and literature. Even Plato told a story involving the finder of a magical object who abused the power that being invisible granted. Scientists believe that full invisibility could be with us within a few decades.

Pluto has a Norse equivalent, appropriately named Hel. She was apparently split into two; the top half of her was living, the bottom half was cadaverous. The Norse goddess Hel supposedly resided in the frozen north, probably the Arctic. The Norse Hades does not seem to be underground, just land in the cold and dark north. After the death of Balder the Beautiful, Odin's favorite son, a god called Hermod rode to Hel. Hermod learned that Balder could leave Hel if everything in the world, both living and dead, would cry for him. As he had been a truly good god who was truly loved, everything and everyone did cry for him, except for one giantess. Because of this one being, Balder could not leave Hel. In some mythologies, this giantess was thought to be Loki, the trickster god in disguise. The goddess Hel is the daughter of Loki!

Unusually for an ancient Greek god Hades/Pluto was faithful, apart from when he fell for a nymph called Minthe. Persephone learned about this and was not going to put up with this. She decided to stop her husband in his tracks by transforming the unlucky nymph into the plant mint.

Pluto's Moons

Pluto has five moons.

Charon

The moon Charon was discovered in 1978 by the American astronomer James Christy. It is almost the same size as Pluto itself. Charon and Pluto always show the same face to each other as Charon orbits Pluto.

Charon ferries dead souls delivered to him by the messenger god Hermes/ Mercury across the river Styx and the other four rivers. The dead person has to pay Charon the ferryman. This money is put in the mouth or on top of each eyelid. If Charon is not paid, the soul will be condemned to linger for a century on the other side of the Styx or Hate. In mythology, Charon was the son of Erebus and Nyx. In some legends, Charon was a centaur who eventually became the boatman, while in others, he was a terrifying ugly old man.

Hydra

Hydra had nine heads that were serpent-like, and one of these heads would live forever. It was a water snake, but with a body that resembled a lion. It was so terrifying and poisonous that even its breath could destroy life. Heracles/Hercules' Second Labor was to destroy this vile creature. This task was almost impossible, as Heracles found out, because when one of the Hydra's serpentine heads was cut off two others grew to quickly replace it. Heracles got Iolaus to help by cauterizing the freshly severed heads with fire so that the blood supply would be cut off and a new head could not grow. The special head was severed and then buried under a large rock. Heracles used its venom to dip his arrows in.

Nyx/Night

This is the third moon of Pluto. Nyx is the most terrifying of all the Greek goddesses. She is often shown in an outfit rather like a modern personification of a witch with a black robe, with stars on it or with black wings, the color black representing death, magic and primeval darkness. She, too, has a chariot with black horses to pull it, rather funereal. Rather than being born she was created from Chaos at the same time as, among others, Erebos. She didn't sleep with any of the gods but had managed to produce several fear-inspiring children, including Nemesis, the Fates, Work and Lies.

A Newly Discovered Fourth Moon

The famous British amateur astronomer, Sir Patrick Moore, suggested the name of Cerberus for this. Cerberus was the three-headed dog that guards the gates of Hades. He was, as expected, somewhat fearsome and terrifying, and in some variations of the same myth, he is described as having not three but fifty heads, is fed uncooked flesh and his body is adorned with venomous serpents. Poison, probably from the snakes on the body of Cerberus, was used by Jason's wife, Medea, to enable his father, Aeson, to live longer. (Jason led the Argonauts, brave men from Argos, to steal the enchanted Golden Fleece.) Heracles is supposed to have dragged the canine monster along with strong metal chains.

Cerberus had to prevent the ghosts of the dead from straying from the Underworld, either by guarding a chasm leading down to Hades or by defending the banks of the river Styx. The Norse equivalent to Cerberus/Kerebos is Garm, the hell hound.

Pluto's two unnamed moons were given names in July 2013. Sir Patrick Moore almost got his way, as P4 was awarded the name Kerebos, the Greek equivalent of Cerberus.

Another suggestion for the name of the moon was Erebos, the mysterious but sinister brother of Nyx/Night. Erebos is either a domain within the Underworld or the place Hades itself, which we might regard as hell.

There is a real life Victorian horror story involving two ships: the HMS *Erebos* and the HMS *Terror*. In 1845, there was a famous British Northwest Passage Expedition that was intended to explore the Arctic. It was comprised of these two ships, whose names, rather like in a Greek tragedy, were practically self-fulfilling prophecies. These two ships were under the supreme command of Sir John Franklin, himself famous, along with 134 men. These ships were state of the art, sturdily built and powered by steam technology. They also boasted luxurious central heating aboard.

The ships set off in May 1845 and then disappeared. They were last seen in July 1845. Apparently, they became trapped in the ice. Franklin himself died on Beechey Island in northern Canada in 1847 (we know this, as he left a note there). Because Franklin was so famous and his wife made a fuss, further expeditions were sent to find out the fate of the 1845 Northwest Passage Expedition. In 1854, the Scottish explorer Dr. John Rae was told by local Inuit that the men left the ships and continued to go on foot, but most died from hunger and the extreme cold. Some had even resorted to cannibalism.

As late as 1986, graves from the Franklin expedition were preserved in the ice, so autopsies of the bodies could be conducted. These revealed that the bodies were full of deadly poisonous lead, which must have leaked from the ships. HMS *Erebos* and HMS *Terror* lived up to their names of hell and terror by poisoning their own crew, slowly and painfully, with some men even eating each other. This truly is the stuff of horror films and suitable for a moon of such a sinister planet.

Ships were given names such as Erebus and Terror in order to try to intimidate their enemies. Mount Erebus, Antarctica, was named after the ship of that unlucky expedition. Both ships were re-fitted for the Franklin expedition. HMS *Terror* was the more elderly vessel, having been originally launched in 1813 after being built in the Davy Shipyard in Topsham, Devon. Its captain on the Franklin expedition was Francis Crozier. The HMS *Erebus* was originally launched in 1826 and was built in Wales. The highly experienced Sir John Franklin himself sailed aboard this ship, having been the commanding officer on two previous Arctic expeditions. Both ships were fitted with steam engines, and iron plating was attached to their hulls. The mission was intended to, among other things, complete a never before attempted crossing of the Northwest Passage.

For a long time, it was believed that the men died from lead poisoning caused by their canned food, until the much later autopsies. Some of the bones from the bodies discovered had cut marks on them, reinforcing the cannibalism idea. In true ghost story tradition, the ships have not been found—yet! And despite the inauspicious reputation, other ships have been named HMS *Erebus* since then. In September 2014 one of these two unlucky ships has been found at last by the Canadians.

Pluto's Fifth Moon

In July 2012, a fifth moon of Pluto was discovered, currently referred to as P5. It is a tiny moon of between 6 and 15 miles (10 and 24 km) and is irregular in shape. The International Astronomical Union has suggested the name Styx for it. Styx is the most important river, as it separates the living from the dead. This is where Charon, the boatman, brought the smoke or shadows of the dead.

These two most recent suggested names are the result of an online poll 'Pluto Rocks' sponsored by SETI (the Search for Extraterrestrial Intelligence). The two names came in second and third in the poll. The name that won first place was Vulcan but was rejected as it was already a name in use in astronomy. (It is the name given to a fictitious planet to try and explain the planet Mercury's weird orbit.) Scientists expect to find more moons that could be discovered when NASA's New Horizons spacecraft flies by in 2015. Perhaps one of these will be awarded the name Erebos?

Afterword

The power of the human imagination, the visionary ideas and the desire to look beyond ourselves, our world, our Solar System and even beyond to our universe is as important as science itself. The ancients looked up at a light-pollution free, clear night sky at the celestial bodies and the patterns their lights made in the darkness above them. Their mythology was their way of trying to make sense of what they did not understand. This same human imagination together with scientific innovations has made possible space travel both manned and unmanned, landing a robotic science laboratory, Curiosity, on the planet Mars and the Voyager spacecraft, which headed out of our Solar System in June 2011, complete with messages for other possible intelligent life. Not so long ago, these achievements, not to mention people on the Moon, would have seemed merely the stuff of science fiction.

Mythology has always been used to try to explain the inexplicable. Earthquakes, for ancient Greeks, were caused by Poseidon, the 'Earth shaker' and god of the seas. For ancient Egyptians, a dragon, an aspect of the god Osiris, caused the river Nile to flood, which made their soil fertile for crops. For the ancient Chinese, the black dragon Gong-Gong is blamed for eratic weather. The ancient Greeks believed that the monster Typhon was buried beneath Mount Etna and that the Titan leader Atlas was turned to stone, becoming the Atlas Mountains.

The Celts explained the Giants Causeway by the myth in which the Irish giant Finn MacCool formed a path for the Scottish giant Bennadonna. The ancient Greeks and Romans looked up at the sky and named the planets after their own gods, names that we still use today and have expanded upon with Uranus, Neptune, Pluto and the moons of planets in our Solar System. They also connected each

© Springer Science+Business Media New York 2015
R. Alexander, *Myths, Symbols and Legends of Solar System Bodies*, The Patrick Moore Practical Astronomy Series, DOI 10.1007/978-1-4614-7067-0

known celestial body with a metal. The Moon was allocated the metal silver, perhaps because it looks silvery. However, in 2010, NASA scientists revealed that Moon dust does actually contain particles of silver, so they were right!

The Man on the Moon began with mythology and ended with reality, with Neil Armstrong and Buzz Aldrin and others after them actually setting foot on our satellite. This was made the subject of mythology very recently when Neil Armstrong died. His family expressed that they hoped that people would think of him when the Moon smiled down at them, making Neil Armstrong the Man on the Moon in people's imaginations.

Mythologies began back in times before writing systems were around, and by their very nature are constantly changing. Much mythology has been lost as the human race has journeyed through time. However, new mythologies, usually based on older tales, emerges sometimes, such as in the aforementioned example of Neil Armstrong. Mythology, it can be argued, is part of history, as it reveals much about human nature and about the mindset of the culture that it originated from. Therefore, Greco-Roman mythology lives on through the names of the planets and their moons, as do Norse and Inuit myths, to a lesser extent entrenching these names and deities in the human psyche for the foreseeable future.

Some myths and legends remain part of the human psyche and imagination, like that of the English King Arthur, whose alleged reputation lives on in the choice of current English royal names. Prince William or the Duke of Cambridge's full name is William Arthur Philip Louis Windsor. Astromythology remains in flags, symbols of entire nations, in language and in divisions of time-lunar months, solar years and festivals such as Christmas derived from Saturnalia.

The future is exciting, with people pushing the boundaries of human endurance, such as the British Ben Saunders and the Frenchman Tarka L'Herpinieres, who recently trekked from Captain Scott's hut 1,795 miles to the South Pole. Space exploration is constantly becoming more exciting, with the spacecraft Voyager traveling into unchartered territory and ESA's Rosetta spacecraft's exploration of a comet, a Chinese manned mission to the Moon and a possible manned mission to Mars in the 2030s and even to an asteroid. Perhaps these future explorers will become immortalized through mythology, maybe through stories and by having Moon craters or features named after them. Maybe we will even discover the answer to that intriguing question one day, "Are we alone?"

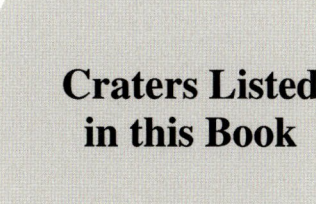

Craters Listed in this Book

Mercury

Apollodorus	Hiroshige	Raphael
Bernini	Homer	Rodin
Cezanne	Kuiper	Shakespeare
Chopin	Michelangelo	Sophocles
Dali	Monet	Sturlusun
Dickens	Mozart	Verdi
Harunobu	Ovid	
Hesiod	Rachmaninoff	

Moon

Aldrin	Flamsteed	Michael
Alexander	Gagarin	Montgolfier
Amundsen	Gauss	Newton
Anaxagoras	Grace	Ohm
Anaximander	Hercules	Osama
Anaximenes	Herschel	Osiris
Anders	Hertzsprung	Pauli
Apollo	Hubble	Peary
Archimedes	Hypatia	Planck
Aristarchus	Ian	Plato
Armstrong	Isabel	Polo
Artemis	Isis	Ptolemaeus
Atlas	Ivan	Ravi
Becquerel	Jose	Rosa
Borman	Kasper	Ruth
Chang'ngo	Kathleen	Samir

(continued)

© Springer Science+Business Media New York 2015
R. Alexander, *Myths, Symbols and Legends of Solar System Bodies*, The Patrick Moore Practical Astronomy Series, DOI 10.1007/978-1-4614-7067-0

(continued)

Chaucer	Kepler	Scott
Collins	Korolov	Shackleton
Colombo	Kuiper	Shahinaz
Daedalus	Lockyer	Shoemaker
Dante	Lovell	Stella
Descartes	Mary	Susan
Dionysus	Matsutov	Taizo
Edith	Mavis	Verne
Einstein	Maxwell	Viviani
Endymion	Melissa	Wells
Eratosthenes	Mendeleev	Yoshi
Euclides	Menelaus	
Euler	Mercurius	
Venus		
Agrippina	Grace	Potter
Akeley	Grey	Quimby
Amelia	Gudrun	Rachel
Bathsheba	Hannah	Rhoda
Beauvoir	Heather	Ruth
Behn	Hiromi	Sarah
Bernice	Irina	Seymour
Boleyn	Isabella	Shakira
Boyd	Jane	Sophia
Bugoslavskya	Johnson	Stuart
Caroline	Juanita	Tako
Christie	Judith	Tamaara
Cleopatra	Kahlo	Ulla
Clio	Karen	Vanessa
Deborah	Katusha	Wazata
Denise	Liv	Whiting
Devorguilla	Macdonald	Witney
Dheepa	Margarita	Wollstonecraft
Elizabeth	Miriam	Xenia
Emma	Morisot	Yoko
Florence	Noriko	Zeinab
Fatima	O'Keeffe	Zoya
Gentileschi	Olga	
Godiva	Pasha	
Mars		
Beagle	Heimdall	Newton
Burroughs	Herschel	Roddenberry
Canaverel	Huggins	Sagan
Columbus	Jodrell	Schiaparelli
Copernicus	Kepler	Stokes
Eagle	Lockyer	Texas
Ejriksson	Lowell	Wells
Endurance	Magellan	Wright
Galilaei	Mariner	
Goldstone	Nereus	

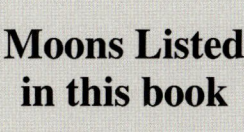

Moons Listed in this book

Earth
Moon
Mars
Deimos
Phobus
Jupiter

Adrastea/Adrestia/Adrastus	Europe	Kore/Core/Semele/Persephone
Aitne	Europie	Leda
Amalthea	Eurydome/Eurynome	Lysithea
Ananke	Ganymede and Harpalyke/Harpalyce	Megaclite and Metis
Aoede	Hegemone	Mneme
Arche	Helike/Helice	Orthosie
Autonoe	Hermippe	Pasiphae
Callisto	Herse	Pasithee/Pasithea
Carme	Himalia	Praxidike
Carpo/Carpho	Io	Sinope
Chaldene	Iocaste/Epicaste	Sponde
Cyllene	Isonoe	Taygete
Elara	Kale/Cale	Thebe
Erinome	Kallichore	Thelxinoe
Euanthe/Euanthes	Kallirrhoe/Callirrhoe/Callirhoe	Themisto
Eukelade	Kalyke/Calyce	Thyone

Saturn

Aegaeon	Hati	Phoebe
Aegir/Eagor	Helene	Polydeuces
Albiorix/Toutatis	Hyperion	Prometheus

(continued)

© Springer Science+Business Media New York 2015

R. Alexander, *Myths, Symbols and Legends of Solar System Bodies*, The Patrick Moore Practical Astronomy Series, DOI 10.1007/978-1-4614-7067-0

(continued)

Anthe	Hyrrokkin and Iapetus	Rhea
Atlas	Ijiraq	Siarnaq
Bebhionn	Janus	Skathi/Skadi/Skade
Bergelmir	Jarnsaxa	Skoll
Bestla	Kari	Surtur/Surt/Surtr
Calypso	Kiviuq/Quviuq/Kiviuk/Qooqa	Suttung
Chiron (Hypothetical moon)	Loge	Tarqeq/Tarquiup
Daphnis	Methone	Tarvos
Dione	Mimas	Telesto
Enceladus	Mundilfari	Tethys
Epimethus	Narvi	Themis (Hypothetical moon)
Erriapus/Erriapo	Paaliaq	Thrymr
Fabauti/Faubauti	Pallene	Titan
Fenrir and Fornjot/Fornjotr	Pan	Ymir
Griep	Pandora	
Uranus		
Ariel	Fransisco	Prospero
Belinda	Juliet	Puck
Bianca	Mab	Rosalind
Caliban	Margaret	Setebos
Cordelia	Miranda	Stephano
Cressida	Oberon	Sycorax
Cupid	Ophelia	Titania
Desdemona	Perdita	Trinculo
Ferdinand	Portia	Umbriel
Neptune		
Despina/Despoena	Naiad	Sao
Galatea	Nereid	Thalassa
Halimede	Neso	Triton
Laomedia	Proteus	
Larissa	Psamathe	
Pluto		
Cerberus/Erebus (suggested names for Pluto's fourth moon).	Hydra	Nyx/Night
Charon	Kerebos	Styx

Bibliography

Prometheus Bound and Other Plays by AESCHYLUS. Translated by Philip WELLA COTT. Penguin Books, London. 1984

The Oresteian Trilogy by AESCHYLUS. Translated by Philip VELLA COTT. Penguin Books, London 1988.

Kingsway Publications, Eastbourne. 1995.

The Life of Alexander the Great by ARRIAN. Translated by Aubrey DE DELINCOURT. Penguin Book, London

Birds of the World by Oliver L. Austin, Jr. Hammlyn, London. 1971.

The Continuum Encyclopaedia of Symbols by Udo BECKER. Continuing Publishing Group, London 2005

Dictionary of Celtic Mythology by Peter BERRESFORD ELLIS. Oxford University Preso, Oxford 1992

Dictionary of Symbolism by Hans BIEDERMANN. Facts on File Inc, New York. 1992

Magic Symbols of the World by Pearl BINDER. The Hamlyn Publishing Group Ltd, Middx, England. 1972

The Illustrated Book of Signs and Symbols by Miranda BRUCE-MITFORD. Dorling Kindersley Ltd, London 1996.

Coll the Storyteller's Tales of Enchantment: The Wild Hunt by Lucy COATS and Anthony Lewis. Orion Books, Orion Publishing Group Ltd, London 2007

Arthurian Legands by Ronan COGHLAN. Elment Books Ltd, Dorset, England. 1991

Brewer's Myth and legend (ed) by J.C COOPER. Cassell Publishers, London 1992.

The Illustrated Encyclopaedia of Myths and Legends by Arthru COTTERELL. Cassell Publishers Ltd, London 1989.

The Encyclopaedia of World Mythology by A. COTTERELL and R. Storm. Anness Publishing Ltd, London. 2007

The Times World Mythology (ed) by Willaim G. Doty. Times Books, London 2002.

Oxford Companion to English Literature (ed) by Margaret DRABBLE. Oxford University Press, Oxford 2000.

Signs and Symbols in Christian Art by George FERGUSON. Oxford University Press, Oxford 1961

© Springer Science+Business Media New York 2015

R. Alexander, *Myths, Symbols and Legends of Solar System Bodies*, The Patrick Moore Practical Astronomy Series, DOI 10.1007/978-1-4614-7067-0

The New Secret language of Symbols by David FONTANA. Watkins Publishing, London. 2010

Symbols of the Goddess by Clare GIBSON. Saraband (Scotland) Ltd., Glasgow 2004

The Handbook of Astronomy by Clare GIBSON. D & S Books Ltd., Devon, England 2005

Wolf: Legend. Enemy. Icon. By Rebecca L. GRAMBO. A & C Black Publishers Ltd., London 2008

Greek Myths by Robert GRAVES (Combined Edition). Penguin Books, London 1992

Theogony and Works and Days by HESIOD. Translated by Dorothea WENDER. Penguin Books, Ltd., 1973

The Poetic Edda. Translated by Lee. M. HOLLANDER. Penguin Books, Ltd., London

The Iliad by HOMER. Translated by Robert FAGLES. Penguin Books, Ltd., London 1991

The Illustrated Odyssey by HOMER. Translated by E. V. RIEU. The Rainbird Publishing Group Ltd.,/Book Club Associates, London 1981

Myths and Legends by Anthony HOROWITZ. Kingfisher Publications Plc., London 1991

The Oxford Companion to Classical Literature (ed) by M.C.HOWATSON. Oxford University Press, Oxford. 1989

Mythology of the Aztecs and Maya by David M. JONES. Anness Publishing Ltd., London 2003

English Fairy Tales and Legends by Rosalind KERVEN. National Trust Books, London. 2008

The Earth's Cycle of Celebration by Glennie KINDRED. Glennie Kindred Publishers, Derbyshire, England 2002

The Curse of the Shamen by Michael KUSUGAK. Harper Trophy, Canada 2006

The Saga of the Volsungs. Translated by Eirikr MAGNUSSON and William MORRIS. Digireads. com.Publishing, Stillwell, KS, 66085 2005

Mythology of the North Indian and Inuit Nations by Brian L. MOLYNEAX. ANNESS Books Ltd, London 2004

Metamorphoses by OVID. Translated by David RAEBURN. Penguin Books Ltd., London 2004

Secrets of the Stone Age by Richard RUDGLEY. The Random House Group Ltd, London 2000

Cassell's Dictionary of Superstitions by David PICKERING. Cassell, London. 1995

The Norton Anthology of English Literature (ed) by Julia REIDHEAD. Seventh edition. Volume 1. W W Norton & Company Ltd., London 2000

The Complete Works of William Shakespeare by William SHAKESPEARE. Wordsworth Editions Ltd., Ware, Hertfordshire, England 1996.

Celtic Mysteries: The Ancient Religion by John SHARKEY. Thames and Hudson, London. 1975;1987

Cosmos by Giles SPARROW Quereus Publishing, London. 2007

Ancient Egyptian Myths and Legends by Lewis SPENCE. Dover Publications Inc., New York. 1990

The Secret Language of the Reniassance by Richard STEMP. Duncan Baird Publishers, London. 2006

Dracula by Bram STOKER. Constable & Robinson Ltd., London. 2012

The Orchard Book of Stories from Ancient Egypt by R. SWINDELLS. Orchard Books, London. 2000

Everyman's Dictionary of Non-Classical Mythology by Egerton SYKES. J.M. Dent and Sons Ltd., London. 1961

Symbols and their Meanings by Jack TRESSIDER. Duncan Baird Publishers, London. 2000

The Watkins Dictionary of Symbols by Jack TRESSIDER. Watkins Publishing, London. 2008

Essential Visual History of World Mythology (ed) by Juliane VON LAFFERT. National Geographic Society, Washington. 2008

The Observer Book of Space (ed) by Carl WILKINSON. Observer Books 2007

Eyewitness Companions: Mythology by P. WILKINSON and N. PHILIP. Dorling Kindersley Ltd.,London. 2007

Lost Planet explains Solar System Puzzle. Article by Lisa GROSSMAN in NewScientist 1/10/11 (pp. 14/15). NewScientist UK London.Reed Business Information Ltd., England.

Websites

About.com
Answers.yahoo.com/question/index%3F
Brkingsolver.com/reference/73-the-s…
Archive.fieldmuseum.org/../delia.html
encyclopaediaBritannicamobile
enchanteddoorway.tripod.com/vamp/
en.inforapid.org/index.php%3Fsearch
en.wikipedia.org
finds.org.uk
idp.bl.uk/../sky.html
islam.about.com/../crescent_moon.html
islam.about.com/../moonsighting
Lloydbleckcollection.cs.uct.ac.za/s
Mapsofworld.com
Onespiritx.tripod.com/gods22.html
Skywalker.cochise.edu/../Vesuvius.html
Smallfarm.about.com/od/sustainablea…
Specialiststudies2.blogspot.com/201
Thecoldestjourney.org
Viking.no/e/people/e.knud.html
Volcano.oregonstate.edu/../japan.html
www.ancientexts.org/../lludd.html
www.angelfire.com/realm/shades/demo
www.angelfire.com/../stories.html
www.answers.com
www.assa.org.au/edm.html
www.bbc.co.uk/../werewolf.shtml
www.bbc.co.uk/news/science-environment-25143861
www.behindthename.com
www.biodynamics.in/Rhythm.html
www.blarneycastle.ie/page/kiss_the
www.britannia.com/../EBchecked/topic/…
www.britannia.com/../kappa
www.celtnet.org.uk
www.chinapage.com/../syho.html
www.chinesefortunecalendar.com/midf
www.cleandungoen.com/../mimung.html
www.distinguishedwomen.com/biograph
www.dragonstrike.com/mrk/myths.html
www.encyclopaedia.com
www.english.pravda.ru/society/anomalousphenomena
www.essortment.com/werewolves_lycon
www.godchecker.com/pantheon/native
www.himalayanvoices.org
www._history.mcs.st_and.uk/Biogra
www.kilaueadventure.com
www.levity.com/alchemy/islam13.html
www.mars.com
www.mysteriousbritain.co.uk/folklor
www.mythencyclopaedia.com

www.mythicalcreaturesguide.com/page
www.nasa.gov/../index.html
www.pantheon.org
www.pilt.edu/~dash/type1335a.html
www.royal.gov.uk
www.sacred-texts.com/../mau18.html
www.sciencehowstuffworks.com
www.theadventureblog.blogspot.com/2012/
www.timelessmyths.com/../beings.html
www.uppsalaonline.com/../valkyrie.html
www.wsd1.org/../maya1.html
Amazon: Unnatural Histories. BBC4 8/5/12
BBC News
Bigfoot Files Channel 4 20/10/13
Britain's Secret Treasures ITV1 19/7/12 with Historian Bettany Hughes
Chemistry. A Volitile History BBC4 2/5/11 with Professor Jim Al-Khalili
Daily Politics BBC2 27/4/12
Greek Myths: Tales of Travelling Heroes. BBC4 15/11/10 with Historian Robin Lane Fox
Horizon Asteroids: The Good, the Bad and the Ugly. BBC4 22/9/12
Iceland: Ash Cloud Apocolypse Channel 5 26/4/13
Illuminations: The Private Life of Medieval Kings BBC4 18/1/12 with Dr Janina Ramirez
Inside the Medieval Mind: Power. BBC4 7/5/12 with Professor Robert Bartlett
Lost Cities of the Ancients. BBC4 24/3/12, 31/3/12.
Pagans and Pilgrims: Britain's Holiest Places: Water. BBC4 18/4/13
Stories from the Dark Earth: Meet the Ancestors Revisited. Sacred women of the Iron Age. BBC4
    9/11/13
The Beauty of Diagrams: Nicolaus Copernicus' theory of a Sun-Centred Universe. BBC4 25/11/10
The Code: Prediction. BBC2 3/8/11, 10/8/11 with Professor Marcus du Sautoy
The Comet's Tale: The Story of Man's Understanding of these Celestial Objects. BBC2 10/3/08
The Mayan Apocalypse. Documentary. Quest. 6/10/12
The Story of Vaisakhi. BBC1 10/4/11
Timewatch: The Secret History of Genghis Khan BBC4 28/12/11
Visions of the Future: The Quantum Revolution. BBC4 23/8/10 with Professor Michio Kaku
Wonders of the Solar System: Empire of the Sun. BBC2 7/3/10 with Professor Brian Cox
Also information from lecture notes from the University of Hertfordshire 1989 and the answers to
    many questions that I have asked to:
Marek Kukula, Public Astronomer at the Royal Maritime Museum, Greenwich
Greg Smye-Rumsby and Elizabeth Cunningham at the Peter Harrison Planetarium, Greenwich.

Index

© Springer Science+Business Media New York 2015
R. Alexander, *Myths, Symbols and Legends of Solar System Bodies*, The Patrick
Moore Practical Astronomy Series, DOI 10.1007/978-1-4614-7067-0

Printed by Printforce, the Netherlands